BRITISH PALAEOZOIC FOSSILS

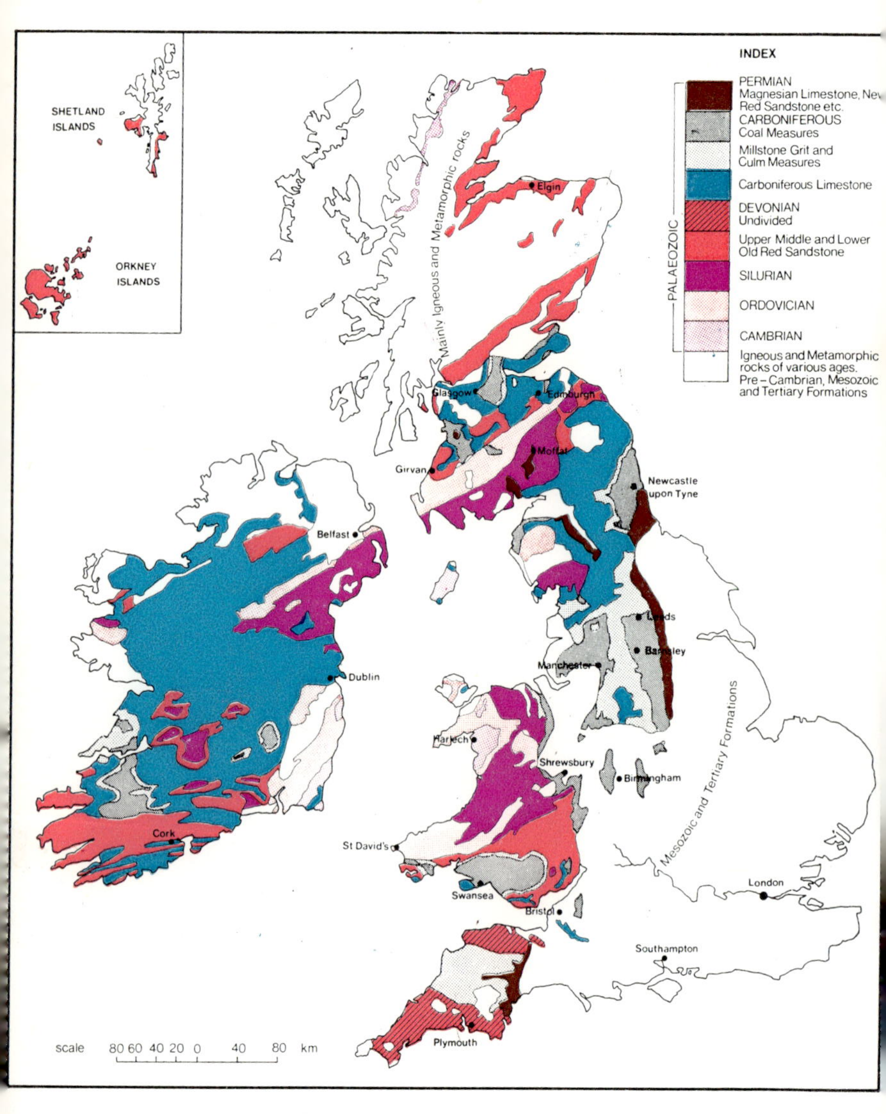

Distribution of Palaeozoic Strata

British Museum (Natural History)

BRITISH PALAEOZOIC FOSSILS

FOURTH EDITION

British Museum (Natural History)
London: 1975

First Edition . . .	1964
Second Edition . . .	1966
Third Edition . . .	1969
Fourth Edition . . .	1975

Publication No. 624

ISBN 0 565 05624 7

BMNH/182/20M/9/75

Printed in England by Staples Printers Limited at The George Press, Kettering Northamptonshire

Contents

Preface

This volume forms part of a trilogy, together with *British Mesozoic Fossils* and *British Caenozoic Fossils*. To quote from the first edition, the aim is to provide 'a series of handbooks illustrating British fossils . . . issued in response to repeated requests for a simple and inexpensive book to enable the young, or those without experience, to know what fossils they may expect to find or, even more important from our point of view, to identify for themselves those they have collected.' Although rising costs have forced the price of all books above the level for which the original compilers had intended, we hope that the aim has been realised, a hope that has been enforced by the encouraging acceptance and sales of the first editions.

Illustrating the life of such an immense period of time as the Palaeozoic Era, encompassing some 350 million years, presents severe problems of selection, not made easier by the frequent distortion and indifferent preservation of many of these ancient fossils. Four hundred and forty-three species are illustrated, which is seventy-eight more than in the Mesozoic volume, but, since the period is so much longer, they represent a smaller proportion of the total number of species known to science. However we have tried to select those animals and plants which are most commonly found in this country, and feel that a useful survey has been achieved, despite the limitations of space.

The series has been produced by the Department of Palaeontology: it was initiated by the late W. N. Edwards, and the first edition brought to fruition by his successor, Dr E. I. White. Each specialist within the Department has been responsible for the selection and identification of the animals or plants within his group. The drawings were chiefly the work of Miss J. C. Webb and Mr D. Erasmus. The introduction and stratigraphical tables, originally prepared by W. T. Dean, have been revised for this edition by L. R. M. Cocks and C. H. C. Brunton.

H. W. BALL
Keeper of Palaeontology

May, 1975

Introduction

The earth is thought to be some 4600 million years old. This vast span of time can be divided into two very unequal units—the Cryptozoic, of broadly 4000 million years duration during which there is only very rare evidence of life, and even then of only the most primitive forms; and the Phanerozoic, embracing the last 570 million years of the earth's history during which there is abundant evidence of life, especially of the higher forms.

The first traces of life occur in rocks that are nearly 3000 million years old. These are primitive single-celled organisms that are comparable with bacteria and blue-green algae. At a much later date, between 600–700 million years ago, fossils of primitive, soft-bodied animals occur in Australia, North America, Russia and at Charnwood in Leicestershire.

However, the first relatively common appearance of fossils marks the beginning of Phanerozoic time, which is subdivided into the Palaeozoic, Mesozoic and Caenozoic Eras. The opening of the Palaeozoic (literally meaning 'ancient life') coincides with a remarkable burst of evolutionary activity some 570 million years ago, and its close 345 million years later (i.e. 225 million years ago) is synchronous with a further period of profound change in the fossil faunas, though not in the floras.

Many Palaeozoic fossils represent animal groups which have long been extinct, and the challenge of deducing their morphology, evolution and ecology has thrilled and taxed many palaeontologists. The study of these fossils has also proved fundamental in our understanding and relative dating of the rocks in which they are found, and in the economic exploration and exploitation of those rocks for coal, oil and other minerals. The Palaeozoic rocks have been subdivided into six systems, the Cambrian, Ordovician, Silurian, Devonian, Carboniferous and the Permian. These six are often formed into two groups of three Systems, termed Lower and Upper Palaeozoic. A summary of their distribution in Britain now follows. For further details the reader's attention is drawn to the series of Regional Guides, each one covering a separate part of Britain in more detail, which are published by the Institute of Geological Sciences.

The whole of Britain west of a line running roughly north-north-east from the Devonshire coast near Exmouth to the coast of County Durham near Hartlepool is composed largely of Palaeozoic rocks (see frontispiece). These strata do not crop out at the surface over the south-eastern half of

England, but their presence in that region has been proved by means of boreholes penetrating the overlying cover of Mesozoic and Caenozoic rocks. The oldest Palaeozoic rocks belong to the Cambrian System, so-called after Cambria, the ancient name for Wales, where they were first studied. There is a fourfold subdivision of the beds into Comley, St Davids, Merioneth and Tremadoc series; and they rest on Pre-Cambrian rocks, from which they are separated by an unconformity or break in deposition, In the Midland and Welsh Border counties Cambrian rocks form fairly restricted outcrops comprising mainly sandy strata, some now compacted into quartzites, laid down near the fringes of a shallow sea. Examples of poorly fossiliferous basal quartzites occur in Salop at The Wrekin, in Herefordshire at the Malvern Hills, and in West Midlands at the Lickey Hills. Farther west, in North and South Wales, the Cambrian rocks are much thicker and represent deposition in deeper water. Fossils are not, on the whole, common, but by means of the trilobite remains the vertical succession of strata has been subdivided into a series of 'Zones' characterized by particular genera and species. The remains of these extinct arthropods outnumber those of all other fossil groups from the Cambrian Era, and the trilobites of the Anglo-Welsh area show close affinities with corresponding faunas in Scandinavia and other parts of Europe. Although many different fossils have been recorded from the Cambrian rocks, specimens are not always easy to find and extract, and especially in Wales they have frequently been distorted by the numerous earth-movements which have taken place since the animals were buried on the sea-floor. The fossiliferous Cambrian horizons in Salop and Warwickshire are not, in general, well exposed, and much of our knowledge of them has been established only as the result of prolonged research and excavation. Fossiliferous Lower Cambrian limestones occur at the tiny village of Comley in Salop, where they are overlain by Middle Cambrian sandstones and Upper Cambrian grey and black shales with brachiopods. A somewhat similar succession is found in the Malvern Hills, where the black shales have yielded trilobites near the village of Whiteleaved (or White Leaf) Oak, whilst in Warwickshire fossiliferous Middle Cambrian shales occur near Nuneaton. Elsewhere, fossiliferous Middle Cambrian rocks may be examined on the coast of South Wales to the east of St David's, Dyfed, where the cliffs at Porth-y-rhaw, near Solva, contain occasional large specimens of the trilobite *Paradoxides*. In the Upper Cambrian, outcrops of the Lingula Flags, so-called on account of the abundant brachiopod *Lingulella* (previously *Lingula*) *davisi*, are widespread and fossiliferous in both South Wales, near St David's and Haverfordwest, and North Wales, near Dolgelly and Portmadoc.

In Scotland the Cambrian rocks form a restricted outcrop confined to the North-West Highlands and only the Lower Cambrian is represented, consisting of a basal quartzite followed by what are known as the Fucoid Beds, dolomitic mudstones and shales with worm tracks. The trilobites obtained from the shales include the genera *Olenellus* and *Olenelloides*, which are unknown from corresponding strata in England and Wales. They apparently inhabited a separate sea which extended westwards to Greenland and eastern North America where similar fossils have been found.

At the top of the Cambrian in England and Wales lie the rocks of the Tremadoc Series, named after Tremadoc in Gwynedd, and there is a local unconformity above these strata in their type-area, with overlying Ordovician beds.

The rocks of the Tremadoc Series in the type area comprise dark slates and mudstones containing trilobites which include the genera *Angelina*, *Asaphellus* and *Shumardia*. Notable localities occur near Portmadoc and Penmorfa, Gwynedd. This series also marks the first known occurrences of the extinct animals known as graptolites. In the rock these appear typically as single- or multi-branched colonies, sometimes replaced by iron pyrites and generally flattened by compaction of the sediments in which they occur. At one time or another they have been linked by palaeontologists with numerous other animal groups, but are at present regarded as having affinities with the hemichordates, a primitive group allied to the vertebrates. There were two main groups of graptolites, first the so-called dendroids, mesh-like conical forms, for example *Dictyonema*, which appeared in the Tremadoc Series and ranged sporadically through to the Carboniferous; and the true graptolites, such as *Didymograptus* and *Monograptus*, which ranged through the Ordovician and most of the Silurian in Britain. The graptolites were marine in habitat and some of the species have been found to be of extremely wide lateral distribution. This, coupled with the fact that they underwent rapid evolution and marked morphological change during their long history, makes them very useful to the geologist as zonal fossils. To the west of Portmadoc there occurs a widely-distributed thin horizon of dark grey shales, weathering brown, known as the *Dictyonema* Band, in which the dendroid graptolite of that name is found abundantly, preserved in iron pyrites. The stratal equivalents of the Tremadoc Slates of North Wales are well represented in the Midlands and Welsh Borders, where they occur in Salop as the Shineton Shales, at Tortworth, Gloucestershire as the Breadstone Shales, in the Malvern district as the Bronsil Shales, and near Nuneaton as the Merevale Shales. These strata form soft, easily-eroded shales and mudstones which contain, especially, graptolites (*Clonograptus*, *Dictyonema*) and trilobites with some cystids (*Macrocystella*) and worms. The exposures in the brook-section at Sheinton (present-day spelling of Shineton) have been a classic collecting ground for many years and contain numerous well-preserved trilobites. Some years ago a bore-hole at Calvert, in Buckinghamshire, showed the existence there of similar rocks underground.

The term Ordovician is inevitably associated in many readers' minds with the well-known controversy involving the eminent geologists Sedgwick and Murchison about the middle of the last century. With Sedgwick working his way upwards through the Cambrian succession of North Wales, and Murchison working his way down through the Silurian succession of the area farther east and south-east, it was almost inevitable that the two should conflict, with the unfortunate result that the same strata were claimed by both geologists as Cambrian or Silurian. Lapworth's compromise solution, whereby he erected a new system, the Ordovician, named after the ancient British tribe of the Ordovices dwelling in the Welsh Borderland, has since found general favour, although occasionally one may still encounter older publications in which only Cambrian and Silurian are used. Although the terms Lower, Middle and Upper Ordovician are sometimes used, it is more usual for geologists to divide the Ordovician rocks into five Series as follows: Arenig (after Arenig Mountain, Gwynedd), Llanvirn (after Llanvirn, Dyfed), Llandeilo (after Llandeilo, now Llandilo, Dyfed), Caradoc (after the Caradoc Hills, near Church Stretton, Salop), and Ashgill (after Ash Gill, near Coniston Water, in the Lake District).

In South Wales rocks of Arenig age occur in the neighbourhood of Whitland and include dark mudstones and shales, some of which contain characteristic graptolites of the genus *Didymograptus*, whilst on Ramsey Island, off St David's, there are outcrops of both graptolitic strata and others, the *Neseuretus* Beds, which contain brachiopods and the trilobite *Neseuretus ramseyensis*. The Llanvirn Series is well developed in South Wales, the type succession being situated near Abereiddy Bay, north-east of St David's. The rocks are black mudstones and shales in which graptolites are not uncommon, and the south side of Abereiddy Bay is well known for the occurrence there of large numbers of the so-called 'tuning-fork' graptolite *Didymograptus murchisoni*, strikingly preserved as white films on a background of dark shale matrix. Elsewhere in South Wales graptolitic shales of Llanvirn age are to be found around the towns of Narberth, St Clears and Whitland as well as Llandrindod Wells farther north. In west Salop the Arenig Series begins with an almost unfossiliferous quartzite, the Stiperstones Quartzite, but this is succeeded by a large thickness of dark mudstones and shales in which graptolites, brachiopods and trilobites occur. Fossils may be collected from several localities in the Minsterley district, whilst the beds cropping out beside the church at the hamlet of Shelve are famous for their graptolite fauna, which includes numerous dendroid forms. The succeeding Llanvirn strata of the district are somewhat more varied in type, including some volcanic rocks, but they contain both graptolites and trilobites and may be seen at Betton Dingle, near Meadowtown. The trilobites in this part of the succession are closely similar to contemporaneous faunas known from Czechoslovakia (so-called Bohemian faunas) and must have lived in a sea connecting both regions.

In North Wales the Arenig Series crops out in the vicinity of Portmadoc and Bangor, Gwynedd. Fossils are not usually abundant but near Arenig Station, Gwynedd, the so-called '*Calymene* Ashes' contain numerous brachiopods and the eponymous trilobite *Neseuretus* (previously *Calymene*) *ramseyensis*. The Skiddaw Slates of the Lake District form extensive out-crops centred roughly on Derwentwater and Skiddaw itself, but the rocks, mainly shales and mudstones, have been extensively folded and cleaved, and fossils are not easily collected. However, numbers of graptolites have been found, some of them in a good state of preservation particularly in the vicinity of Keswick and Mungrisdale, indicating that most of the succession belongs to the Arenig and part of the Llanvirn Series, whilst the Tremadoc may also be represented in the lower strata.

We noted earlier that the Lower Cambrian fossils of the North-West Highlands are of eastern North American type. The Lower Cambrian beds are there succeeded by a poorly fossiliferous series, the Durness Limestone. The few fossils known from the latter include trilobites, gastropods and cephalopods. They indicate that the affinities of the lime-stone lie with the Canadian Series (late Cambrian and early Ordovician) of Greenland and eastern North America, a relationship which was paralleled again, during the Middle and Upper Ordovician, in the Girvan district of the Southern Uplands.

The Llandeilo Series is typified by a fossiliferous sequence of beds near the town of Llandilo (modern spelling of Llandeilo) in Dyfed. The Series has been subdivided on the basis of the contained trilobites, in particular the trinucleid genus *Marrolithus*. Characteristic fossils may be collected from mudstones and impure limestone bands at quarries in Dynevor Park, as well as at Llan Mill, near Narberth. Elsewhere in South and Central Wales the Llandeilo Series crops out around the spa towns of Builth Wells and Llandrindod Wells in Powys, the rocks forming a succession of dark shales and mudstones in which graptolites (*Diplograptus*, *Glyptograptus*) and trilobites (*Cnemidopyge*, *Ogygiocarella* and *Trinucleus*) are often abundant. Farther north fossiliferous Llandeilo strata of generally similar type may be examined at the village of Meadowtown in west Salop where they chiefly contain trilobites with subsidiary graptolites and brachiopods. There are restricted outcrops of Llandeilo strata containing characteristic fossils in the Berwyn Hills, in the general area of Llanrhaiadr-ym-Mochnant, Clwyd. These, however, are some of the most northerly occurrences of Llandeilo fossils in England and Wales, the rocks of this age being either absent or represented by the great thickness of igneous rocks forming the Borrowdale Volcanic Series of the Lake District.

The transition from the Llandeilo to the Caradoc Series is continuous in several areas. However, in the area south-east of the Pre-Cambrian and Cambrian rocks forming the Church Stretton Hills in South Salop the only Ordovician rocks developed are those of Caradoc age, but a few miles farther west, in the Shelve area of West Salop, there is a remarkably full development from the Tremadoc to the Caradoc. The Caradoc strata of south-east Salop are famous for their fossils, and the rocks were laid down on the bottom of a shallow sea in which brachiopods, trilobites, bryozoans and ostracods flourished in large numbers. The basal beds are said to be diachronous, that is to say they were not all formed at the same time, owing to the gradual encroachment of the sea over the shoreline of earlier rocks. The valley of the River Onny cuts through almost the whole Caradoc succession of the district, and the highest beds yield large numbers of the trinucleid trilobite *Onnia* at the historically important section in the river bank south-west of the village of Wistanstow. Outside Salop, fossiliferous Caradoc strata are developed over much of North Wales, in particular the Berwyn Hills, the Lleyn Peninsula around Pwllheli, the Bala district and the neighbourhood of Welshpool. The last includes the famous 'Trilobite Dingle', more properly Bron-y-Buckley Wood, with its large fauna of trinucleid trilobites, especially *Salterolithus caractaci*. The igneous rocks of Snowdonia belong to the Caradoc series, and they too sometimes contain shales and mudstones with trilobites and brachiopods, for example at the summit of Snowdon. The higher Caradoc beds of North Wales are graptolitic shales, whilst the whole succession in South Wales is made up almost entirely of dark mudstones and shales in which graptolites are the chief fossils, though an impure limestone with corals occurs at Robeston Wathen, Dyfed. The North of England has fossiliferous Caradoc beds well exposed near the villages of Knock and Dufton in Cumbria, and a narrow outcrop of strata runs diagonally across the Lake District from near Shap Fell in the north-east to Millom in the south-west. The latter rocks yield poorly preserved trilobites, brachiopods and corals at several localities and are overlain by the beds of the Ashgill Series, here forming a parallel outcrop. In Scotland Caradoc beds are developed in the Girvan and Moffat districts of the Southern Uplands. The Moffat succession is made up of graptolitic shales containing genera and species which are found in the Anglo-Welsh area as well as elsewhere, but near Girvan the beds, which are often highly fossiliferous at Balclatchie, Dow Hill and Ardmillan, contain trilobite-brachiopod faunas similar to those of the eastern United States, and bearing little close relationship to contemporary Anglo-Welsh faunas.

The Ashgill Series has its type area near Cautley and fossils may be found at several points along the Lake District outcrop, particularly near the village of Torver, Cumbria. Elsewhere in Northern England characteristic faunas can be collected at the hamlet of Keisley, south-east of Dufton, Cumbria. Farther north the Girvan district includes several prolific localities, one of the best known being Thraive Glen, where the large fauna includes trilobites, brachiopods and echinoderms. Particularly noteworthy are the strange echinoderm-like genera like *Cothurnocystis* and *Scotinecystis* which have gill-slits like a fish and probably represent the group from which vertebrates evolved (calcichordates or carpoid echinoderms). In North Wales fossiliferous Ashgill rocks occur at numerous places, including Rhiwlas, near Bala, Gwynedd; Bryn Hendre and Deganwy, near Conway, Gwynedd; and various points around the Berwyn Hills, such as Glyn Ceiriog. In South Wales they are well exposed around St Clears and Haverfordwest in the Slade and Redhill mudstones. The faunas of shelly type in the Ashgill rocks were more cosmopolitan in character than those of preceding series, and broadly similar assemblages are found not only over the Anglo-Welsh area but also in Scotland, Ireland, Scandinavia, Poland and Czechoslovakia.

Before leaving the Ordovician, mention must be made of an interesting horizon which, although strictly speaking, should be included in the Trias, nevertheless belongs with the Ordovician, and to a lesser extent Devonian, as far as its fossils are concerned. Exposed in the cliff-sections a short distance west of Budleigh Salterton, on the Devon coast, is a bed containing rounded pebbles of liver coloured quartzite. Some of these contain well-preserved trilobites and brachiopods belonging to species otherwise unknown in Britain. The fossils and their enclosing matrix can be matched exactly with those of an horizon in Normandy known as the Grès de May, and the Budleigh Salterton pebbles represent the products of erosion of a land-mass lying south of Britain during the Triassic period. Similar pebbles have also been recorded from the Bunter Sandstone (Trias) in the Midlands, around Birmingham.

The Silurian System was first named by Murchison in 1835 after the Silures, an ancient British tribe inhabiting what is now the Welsh Border region. The system is perhaps best known from its development in the agricultural country of Salop south-east of The Longmynd and the Church Stretton Hills, where the rock-succession is composed mainly of thick mudstones and shales alternating with extensive developments of massive and nodular limestones. The latter are more resistant to erosion than the intervening shales and so give rise to the topographical ridges or 'edges'—Wenlock Edge is the best-known—for which the county is famous.

Over much of the Anglo-Welsh region the time interval covering the later part of the Ordovician and the beginning of the Silurian coincided with pronounced earth-movements—the so-called Taconian Orogeny—as a result of which the highest Ordovician and lowest Silurian strata are absent from some districts. Just as the Ordovician rocks are subdivided into a number of Series named after places in the Anglo-Welsh region, so the Silurian rocks are divided into four Series, namely Llandovery, Wenlock, Ludlow and Pridoli. The lower three series have their type areas in Wales and England, the upper one takes its name from a town in Czechoslovakia, near where the international Siluro-Devonian boundary has been placed.

A complete succession of Llandovery strata occurs in relatively few parts of the British Silurian outcrop. In the neighbourhood of the towns of Llandovery and Haverfordwest in South Wales the earlier Llandovery beds comprise in the main sandy mudstones containing large numbers of brachiopods and some trilobites. They are well exposed along the valley of the River Cleddau near the Gasworks at Haverfordwest, Dyfed. In the Southern Uplands of Scotland mudstones and sandstones of Llandovery age, often highly fossiliferous, occur around Newlands and Mulloch Hill, north-east of Girvan. All these strata represent deposits formed in shallower parts of the sea, but the rocks laid down in the deeper waters of the same sea are of a different facies, comprising a thinner succession of dark shales and mudstones with graptolites (especially *Monograptus*), often abundant and beautifully preserved in iron pyrites. Graptolitic beds of this type are to be found in North and Central Wales, and in the Lake District, where good specimens of *Monograptus* may be collected at Skelgill and Browgill, both near Ambleside, as well as from other parts of the outcrop. The Llandovery beds of the Moffat district, Dumfriesshire, in particular at Dobbs Linn, are also essentially graptolitic, contrasting with the shelly development of Girvan and so following the general pattern of distribution existing in the Southern Uplands during the later Ordovician.

Other deeper water Lower Silurian deposits include very thick, relatively unfossiliferous deposits such as the Aberystwyth Grits, which were laid down in ocean basins by successive turbidity currents.

Outcrops of the highest Llandovery strata are more widely distributed than those of the earlier beds and often highly fossiliferous, containing numerous brachiopods, trilobites and corals. In South Salop there is a development of conglomerates and limestones with the brachiopod *Pentamerus*, resting unconformably on Pre-Cambrian rocks around the margins of The Longmynd. The beds were evidently deposited very close to the shore-line of a sea in which The Longmynd was either an island or an area of very shallow water. Farther south, fossiliferous beds of both similar and later age occur at May Hill, near Longhope, Herefordshire; at Tortworth, Gloucestershire, where the solitary button-like coral *Palaeocyclus* is abundant; and in South Wales at Marloes Bay, Dyfed. In the Welsh Borders the highest Llandovery strata are shales which pass upwards into the rocks of the Wenlock Series, the latter comprising a lower group of shales, followed by a thick group of massive and nodular limestones. The two horizons, the Wenlock Shales and the Much Wenlock Limestone, are typically developed around Much Wenlock. Parts of the Much Wenlock Limestone represent a development of reefs, and large quarries in the vicinity of Much Wenlock and along Wenlock Edge yield especially corals, bryozoans, and brachiopods. To the east, the Wenlock limestone of Dudley was a classic source of fossils and large numbers of beautifully preserved brachiopods, trilobites, corals, bryozoans, echinoderms and other groups were collected during the 18th and 19th centuries, particularly from the outcrops at Wren's Nest Hill. Unfortunately for present-day collectors most of the fossiliferous limestone was removed by quarrying and used as a flux in the smelting of local Coal Measure ironstones, but some fossils are still obtainable from the adjacent strata. As one moves west from Wenlock Edge the Much Wenlock Limestone dies out and is replaced by a more graptolitic development of strata, for example at the Long Mountain, near Middletown, Salop, although an interesting reef limestone of slightly earlier geological age is found near Old Radnor. The graptolites from the Wenlock Series include species such as *Monograptus priodon*, in which the thecae are of conspicuously hook-like form. Generally speaking the Wenlock faunas are widely distributed, and often have marked affinities with their Scandinavian contemporaries. Rocks and fossils of this series have been proved at depth in boreholes in the south-east of England, whilst in parts of North Wales, as well as the Lake District, the rock succession contains coarser sediments in which fossils are less common. Wenlock strata are present in the Southern Uplands of Scotland but the beds are highly folded, and the fossils are mainly graptolites with smaller numbers of trilobites and brachiopods.

The rocks of the Ludlow Series are best known from their type-area around the old town of the same name in south Salop. The succession there, which follows upon the Wenlock Limestone, was at one time divided simply into three parts, the Lower Ludlow Shales, Aymestry Limestone and Upper Ludlow Shales, but these have been replaced by the terms Elton Beds, Bringewood Beds, Leintwardine Beds and Whitcliffe Beds. The Lower Ludlow Shales are soft, graptolitic shales, easily eroded and form the valley of Hope Dale; good exposures are not plentiful, but fossils can be collected at Upper Millichope, five miles south-east of Church Stretton. The succeeding Aymestry Limestone forms the well-known scarp of View Edge, and its north-easterly extension parallel to Wenlock Edge. The limestone is often highly fossiliferous, containing large numbers of brachiopods and corals, but its most spectacular development is seen in the large quarry at Weo Edge, overlooking Craven Arms, where it is composed almost entirely of the valves of the large brachiopod *Kirkidium knighti*, though whole specimens are rather difficult to extract. The shaly strata about the horizon of the Aymestry Limestone contain the last-known British graptolites, though the group survived into the Devonian elsewhere. The succeeding beds, previously known as the Upper Ludlow Shales, are often rich in shelly fossils, particularly brachiopods such as *Dayia* and *Protochonetes*. The rocks may be examined at numerous points along the dip-slope behind View Edge; by the road from Craven Arms to Ludlow; at Ludlow itself, especially near the Castle; and at localities around Bringewood Chase, near Ludlow. The Ludlow rocks of the Welsh Borders and adjacent districts were laid down in the shallower, marginal or 'shelf' areas of the Silurian sea, whilst farther west, in Central Wales, the rocks represent deposition under deeper marine conditions, the so-called 'basin' deposits. The fauna of the basin rocks differs in many respects from that of the shelf deposits and contains more graptolites and molluscs. Limestone horizons are generally absent and the strata, which are considerably thicker than those of the border counties, consist mainly of siltstones and flags, but despite the differences in lithology enough characteristic fossils have been found to effect a reasonable correlation. Ludlow strata of this type form large outcrops in Wales, but particular mention may be made of the areas around Usk and Builth Wells. A conspicuous feature of the rocks both here and in Clwyd, North Wales is the presence of curious contorted strata, and these are interpreted as having been formed by the sliding and slumping of layers of unconsolidated muds along gentle slopes of the sea-floors. Farther north, in the Lake

District, a large thickness of Ludlow strata is developed, forming a wide outcrop extending north-eastwards across the area, parallel to those of earlier strata. The rocks are exposed in the neighbourhood of Windermere, Kendal and Kirkby Lonsdale, and comprise slates, grits and flags in which fossils, particularly brachiopods similar to those of the Welsh Borders, occur, usually in bands.

Over most of the British Isles the period of time about the end of the Silurian and the beginning of the Devonian was marked by pronounced physiographical changes which were reflected in the types of sediments laid down. Earth-movements, forming part of what has been termed the Caledonian Orogeny, caused uplift of the region now embracing Wales and neighbouring areas together with parts of Scotland. As a result, the Silurian marine deposits of Wales and the English Midlands were supplanted by either terrestrial strata or beds laid down under deltaic or semi-fluviatile conditions. With these changes in lithology the marine animals of the Ludlow Series were replaced by assemblages composed essentially of inarticulate brachiopods, gastropods and primitive fish-like ostracoderms which probably lived in fresh and brackish water. The deposits so formed comprise large thicknesses of mainly arenaceous strata to which the name Old Red Sandstone was applied last century by Murchison to distinguish the beds from the later New Red Sandstone, of Permian and Triassic age. In the absence of the more useful marine invertebrate fossils the stratigraphy of the Lower Old Red Sandstone is often difficult to elucidate although primitive plant spores are now being used. The succession has been divided into a number of local Stages and Zones, named after places in the Welsh Borders and characterized by different genera and species of ostracoderms which have proved of great value in correlating the rocks not only in Great Britain but also with other parts of Europe and as far afield as Greenland and North America. These animals were not true fishes for they had no jaws; they carried an armour-like covering of thick plates, and were distantly related to the present-day lampreys and hag-fishes. In the Midland Valley of Scotland the transition from marine to freshwater conditions of deposition began earlier than in the Anglo-Welsh area, and near the town of Lesmahagow, in Lanarkshire, strata of Upper Silurian age are of shallow-water type, passing upwards into typical Old Red Sandstone, and have yielded ostracoderms as well as numerous arthropods, including the eurypterid *Pterygotus*. In parts of the Welsh Borders the base of the Old Red Sandstone is marked by an interesting, though irregularly distributed, deposit known as the Ludlow Bone Bed. As the name implies, it consists essentially of fish fragments and teeth, embedded in a loose, weathered matrix and showing signs of having been rolled around and abraded on the floor of a shallow sea. The bed crops out at Ludlow and at intervals along the north-west side of Corve Dale, a valley running north-eastwards from near Ludlow. Other localities for collecting Old Red Sandstone fishes are distributed over the Clee Hills of Salop, though specimens are not common, whilst across the border in Wales, large outcrops of Old Red Sandstone form the hills of the Brecon Beacons. *Cephalaspis* and *Pteraspis* are perhaps the best-known genera of Old Red Sandstone ostracoderms, whilst the horny brachiopod *Lingula* and the large bivalve *Archanodon* may be locally abundant. In Cornwall the corresponding beds, which are well exposed in Bude Bay and on the south coast around Polperro, have been highly distorted by earth-movements and are known as the Dartmouth Slates. The ostracoderms in them occur in an extremely tattered condition and their fragments were originally described as sponges!

Only traces of Old Red Sandstone occur in the north of England, but the series is well represented in Scotland, and Dura Den, near St Andrews, is a classic locality in the Upper Old Red Sandstone once noted for fossil fishes in an excellent state of preservation, representing the inhabitants of a small lake which had quickly become filled up. The Orkney and Shetland Isles, together with the neighbouring parts of the mainland, also afford good outcrops of Old Red Sandstone. Only the middle and upper parts of the series are represented, comprising mainly sandy beds showing shallow-water characteristics, and the remains of fossil fishes may be locally abundant, for example at Thurso and other places in Caithness. These were true fishes, related to the living lung-fishes and coelacanth.

At the same time as most of the Old Red Sandstone was being laid down in England and Wales, the south-west of England, including Devonshire which gave its name to the Devonian System, was covered by a sea which extended eastwards into Europe, in particular Belgium, north France and Germany. In south Devonshire, limestones are well exposed along and near the coast in the neighbourhood of Torquay, for example at Hope's Nose, and contain numerous corals, brachiopods and trilobites whose affinities are with corresponding faunas in Germany. Some of the Devonian strata of south Devonshire include thin limestones together with other beds indicating muddier conditions of deposition, and their fauna, whilst lacking corals and certain brachiopods, contains the cephalopods known as goniatites, and some trilobites. The former group, so-named after the angular form of the septal sutures—the lines marking the junction of the shell with the septa dividing its interior into separate chambers—bear the same relation to Upper Palaeozoic stratigraphy as do the ammonites in the Mesozoic, and have enabled the rock-succession to be divided into numerous stages and zones which may often be correlated over large distances. Fossil localities are numerous in South Devonshire, but we may perhaps single out the Chudleigh, Newton Abbot and Torquay districts, whilst Mudstone Bay and Saltern Cove, both near Brixham, are well known for their goniatites. Outcrops of marine Devonian rocks extend also into Cornwall, where fossils, including goniatites, have been found at several localities around Padstow. The Devonian strata of north Devonshire represent an intermediate type of succession comprising marine rocks with alternations of Old Red Sandstone type, sometimes with fish and plant remains. Some of the marine strata in the Ilfracombe district contain brachiopods (*Stringocephalus*) and corals (*Heliolites*) whilst at Fremington the Pilton Beds have yielded trilobites, brachiopods and goniatites, and represent a process of sedimentation which was continuous during late Devonian and early Carboniferous times.

The Carboniferous System derives its name from the fact that it contains all the important coal-producing strata of Britain. Despite this, however, there is a great variety of rock-types involved, ranging from pure limestones to sandstones and coals. The thickness of rocks is considerable, and in North America the equivalent strata have acquired the status of two separate systems, the Mississippian and Pennsylvanian. The British Carboniferous succession may be divided approximately into three parts, to which the somewhat generalized names of Carboniferous Limestone, Millstone Grit and Coal Measures, indicative of the typical lithologies, have been applied in ascending order. The Carboniferous Limestone corresponds broadly to the Lower Carboniferous, whilst the Millstone Grit and Coal Measures constitute the Upper Carboniferous, though problems have sometimes arisen regarding the exact position of the boundaries. As in the case of other systems, the Carboniferous rocks have been subdivided into series and zones according to the fossils they contain. The series names are derived from regions in Belgium and Germany where rocks of their particular ages are fully represented.

Rocks of Lower Carboniferous (Dinantian) age are widely distributed over much of the British Isles and in places reach several thousands of feet in thickness. They are commonly grey limestones and shales and their original sediments were deposited in seas which covered most of the British Isles and much of Europe at that time. Lateral changes in rock type reflect changes in the depositional environment and hence in the fossils now to be found in the rocks. Similarly environmental changes through time, especially the deepening and shallowing of the seas, are reflected in the rocks seen today and allow the recognition of major cycles of sedimentation which are used in correlating Lower Carboniferous strata (see page 26). Many localities within the Lower Carboniferous limestones and shales provide prolific fossil collecting; brachiopods, corals, bryozoa, molluscs and crinoids occur in profusion and the remains of trilobites, ostracods, echinoderms, foraminifera, fish and algae may also be found. Of the macrofossils brachiopods, corals and cephalopod molluscs are especially important stratigraphically whilst foraminifera, conodonts (which remain of unknown affinity) and ostracods are useful microfossils.

Probably the most famous of all British Lower Carboniferous sections is that exposed along the gorge of the River Avon at Bristol. The lowest Carboniferous strata, which overlie Old Red Sandstone, crop out at the Avonmouth end of the gorge, and one moves through successively higher beds towards the Clifton Suspension Bridge. At the beginning of the century the area provided the setting for Vaughan's classic work on the Lower Carboniferous succession, which he subdivided into a number of zones, founded on coral and braciopod genera. Although Vaughan's zones are still used and can, with modification, be applied throughout much of Britain and elsewhere, recent work shows that parts of the full Lower Carboniferous succession seen in northern England are missing in the Bristol region which, therefore, has diminished in geological importance. The massive limestones of the Avon Gorge have been extensively quarried, so that exposures are numerous. Characteristic fossils, especially corals and brachiopods were formerly abundant, though their extraction from the fresh rock is now often difficult. The lower Carboniferous outcrop continues south-westwards into the Mendip Hills where the strata are overlain by Jurassic beds, the normally intervening strata being absent. Exposures of fossiliferous limestone are abundant, but particularly fine sections may be examined along the Cheddar Gorge and at Burrington Coombe nearby. West and north of the Bristol district the thickness of Lower Carboniferous rocks gradually diminishes, and in the Forest of Dean, in Gloucestershire, as well as on the eastern side of the Clee Hills of Salop, the rocks are thin and more arenaceous in character, indicating that they were laid down under shallower marine conditions along the margins of a land mass, to which the name St George's Land has been given, comprising most of what is now Wales and extending westwards as far as south-eastern Ireland. In the Forest of Dean, fossils may be found near Coleford and Mitcheldean, whilst at the Clee Hills, sandy limestones in Oreton quarries, near Farlow, yield many brachiopods as well as fish fragments, chiefly the teeth and spines of primitive sharks. In South Wales, Lower Carboniferous limestones and sandstones form a rim enclosing the area of the South Wales Coalfield, but are not encountered again in the Principality until one reaches North Wales, where extensive outcrops of fossiliferous limestones occur in the vicinity of Llangollen, Llandudno and the Vale of Clwyd, and are represented to a lesser extent farther west, on Anglesey. These North Welsh outcrops extend eastwards into the North Midlands, where they are well developed in Derbyshire, especially around the towns of Matlock and Castleton. The rocks here consist mainly of limestones with shelly fossils and subsidiary shales containing goniatites, and represent marine strata deposited on the north and east sides of St George's Land. Among the many interesting features of the Derbyshire strata are the existence of local volcanic rocks, known as toadstones, and the presence in the Lower Carboniferous rocks of some of the few commercial oil deposits in Britain. In the Castleton area there occurs pure but poorly-bedded limestones, sometimes forming steep-sided hills, in which locally abundant faunas of brachiopods and molluscs are to be found. These structures have been termed 'knoll-reefs' and they probably formed in shallow seas, not far from a coastline, in a position somewhat analagous to many present day reefs. Corals, with the exception of the genus *Amplexus*, are generally absent from the knoll-reefs. Travelling north from Derbyshire one does not encounter Lower Carboniferous outcrops again until one reaches the Pennines in the region of Skipton, whence they continue unbroken into Berwickshire. Over the southern part of these outcrops the Lower Carboniferous consists largely of massive grey limestones which form well marked topographical features and abundantly justify the name Mountain Limestone which was applied to them during the last century. However, the details of the rock-succession differ considerably, as do the faunas, from those farther south, and there are also marked lateral variations. Dark shales containing the stratigraphically important goniatites are well represented, particularly in the Bowland (or Bolland) Shales of the Bowland Forest area on the Yorkshire–Lancashire border, whilst knoll-reefs are an important feature of the geology of the Clitheroe–Settle district, where they form conspicuous topographical features. As in Derbyshire, the knoll-reefs are highly fossiliferous, for example at Coplow Quarry near Clitheroe. The reefs are believed to vary in their mode of formation, but their origin is still in question; some believe they represent original mounds of carbonate mud on the sea floor which were surrounded by contemporaneous and younger shales, whilst others believe them to have been sculpted by erosion from pre-existing strata and that the shales filled the spaces between the 'reefs' later.

The highest Lower Carboniferous rocks of the Northern Pennines have been called the Yoredales after the old name for Wensleydale. The rocks represent deposition under a particular set of conditions rather than a definitely restricted horizon, and their age is somewhat variable, so that they sometimes bridge the boundary between Lower and Upper Carboniferous. One of the most interesting features of the Yoredale rocks is the manner in which a particular cycle of deposition is repeated several times, a process known as rhythmic sedimentation. In this case the cycle consists of the ascending sequence of limestone, shale, sandstone and coal, and indicates initial deposition under marine conditions, followed by shallowing so as to produce conditions suitable for plant life which provided the raw material for the coal. Fossiliferous limestones of this group are well exposed in the valleys of Wensleydale and Swaledale, as well as on the neighbouring high ground. When traced into the Bowland Forest area the beds of the Yoredale Series are found to pass laterally into the Bowland Shales. Lower Carboniferous rocks are well developed in the north of England, where good sections are exposed along the Northumberland coast, near Alnwick and Berwick, as well as in the Midland Valley of Scotland. Generally speaking the succession, which is often highly fossiliferous, differs from that found elsewhere in the British Lower Carboniferous in that it includes numerous thin limestone horizons. The strata of the Midland Valley are economically important and contain the so-called Oil Shales, from which crude oil is obtained by distillation, as well as commercial coal deposits. Additional coals occur higher in the Carboniferous succession, coeval with certain of the Anglo-Welsh Coal Measures, of Westphalian age (see Tables), whilst Central and Southern Scotland were the scene of intense volcanic activity throughout much of the Carboniferous period.

The rocks of the Millstone Grit are typically developed in the north of England as a series of alternating sandstones or 'grits' and shales, the former giving rise to the characteristic escarpments west and north-west of Bradford and Leeds, as well as in parts of Derbyshire. The sandstones were deposited in shallow water, in the delta of a large river flowing from a land-mass to the north or north-east. On the other hand, the shales represent deposits of marine muds laid down when the area was periodically submerged below sea-level and they contain the fossils of marine animals, in particular those of the goniatites (e.g. *Reticuloceras*) which, as in the Devonian rocks, have proved of great value in correlation over wide areas. Sometimes the goniatites are preserved as three-dimensional moulds in nodules, but they are more commonly found as flattened impressions in black shales, though the details of the surface ornamentation are often well preserved. Fossils can be found in Derbyshire, at Edale and near Castleton; in Lancashire, near Preston and Whalley;

and in Yorkshire, near Hebden Bridge, Keighley, Lothersdale, Marsden and Todmorden. Strata of Millstone Grit age are well developed around the margins of the South Wales coalfield, where they are separated by a marked stratigraphical break from the underlying Lower Carboniferous rocks.

From an economic standpoint the Coal Measures form the most important part of the geological column in Britain. The coals themselves form only a small fraction of the total thickness of rock present, and these formations represent the fossilized remains of deltas and forest swamps along the fringes of land areas occupying what is now the Highlands of Scotland and also extending across Wales, central England and eastwards into France. Similar conditions prevailed over much of north-east Europe, but in the Mediterranean region and eastern Europe the equivalent strata are marine deposits. The forests of the Coal Measure swamps probably thrived in a warm, moist climate, and several plant groups were represented. They included giant club-mosses or Lycopodiales such as *Lepidodendron* and *Sigillaria*, together with horse-tails or Equisetales (*Annularia* and *Calamites*), tree-ferns or Filicales, and a group of seed-bearing fern-like plants, the Pteridosperms, exemplified by *Neuropteris*. Attempts have been made to subdivide the rocks stratigraphically, on the basis of their included fossil plants, into a number of stages and 'floral zones'; these may be used fairly successfully over short distances, but have generally proved less reliable for correlation than fossil invertebrates. As might be expected the Coal Measure forests supported an insect population, the remains of which, including large dragonflies, have sometimes been found. Vertebrates were also present, and included fishes and amphibians together with the first representatives of the reptiles. The streams traversing the swamps often contained large numbers of bivalves which have been likened to present-day fresh-water forms such as *Unio*. The Coal Measure bivalves, which include the genera *Carbonicola*, *Anthraconaia* and *Naiadites* amongst others, lived in water which need not necessarily have been fresh, and they were probably able to tolerate a certain degree of salinity. Their remains have proved of great value in correlation, and the Coal Measure succession has been divided into zones on the basis of the various genera and species. The swamps were low-lying and underwent occasional submergence, when marine strata were deposited. The latter, known as Marine Bands, are relatively thin by comparison with the other beds but contain marine fossils, for example brachiopods together with goniatites such as *Gastrioceras* which enable more precise correlations to be made. Sometimes the goniatites are found uncrushed, preserved in nodules which are known to the miners as bullions. The fauna of the marine bands may also include fish remains, whilst horizons containing the brachiopod *Lingula* suggest shallow, near-shore conditions of deposition.

As a result of extensive earth-movements late in the Carboniferous period, the rocks formed during that time were folded into a number of troughs and basins which form the principal coal-fields at the present day. Open-cast workings afford facilities for collecting fossils in place, but as coal is mostly mined in Britain it is to the tip-heaps of waste material that the collector must turn for most of his specimens. Exploitation of coal resources has been carried out on such a large scale that it is impossible to give here a comprehensive list of fossil localities, but in the Bristol-Somerset coalfield the Radstock district is well-known for its fossil plants, whilst in the South Wales coalfield the districts of Aberdare, Caerphilly, Merthyr Tydfil and Pontypridd, amongst many others, have yielded abundant floras and faunas. The Coal Measures of Coseley in Staffordshire, and Coalbrookdale, near Ironbridge in Salop, have long been famous for their beautifully preserved fossils, including brachiopods (*Brachythyris*) and arthropods (*Euproops*) as well as plants. In the northern counties fossils of both marine and non-marine type may be obtained in Yorkshire at Baildon, near Bradford, and Horsforth, near Leeds; in Lancashire, near Bolton and Wigan; and in Staffordshire near Burton-upon-Trent. As yet we have made no mention of the Carboniferous rocks in the far south-west of England. The entire Carboniferous succession in Devonshire, a structurally complicated region, is made up of cherts and shales together with some bands of poor quality coal known locally as culm, hence the name Culm Measures which is sometimes applied to the rocks. Though much of the succession is unfossiliferous, some of the beds contain goniatites, bivalves and trilobites which have facilitated a correlation with corresponding strata in Northern England and North Germany. Lower Carboniferous strata of this type crop out to the south-west of Barnstaple, where they include the well-known fossil locality of Coddon (or Codden) Hill, and farther south-west, in the Bideford region, they pass upwards into Upper Carboniferous beds containing occasional plant remains.

We have already noted the extensive earth movements towards the end of the Carboniferous Period in Britain. In consequence of these, non-marine conditions prevailed over the region and the rocks laid down comprise marls and sandstones containing rare organic remains. Thus began the prolonged arid or semi-arid conditions which continued through the Permian and into the Triassic period of the Mesozoic and produced the deposits known as the New Red Sandstone, a series of events analogous in some ways to those which led to the formation of the

Old Red Sandstone some 170 million years earlier. Although the British Isles are well endowed as far as outcrops of Carboniferous and earlier strata are concerned, Permian rocks are comparatively poorly represented and those available afford limited opportunities for the fossil collector, though they are interesting geologically and often important economically. The most fossiliferous British Permain strata are found in the north of England and include, in particular, the Upper Permian of County Durham, where limestones are extensive. The older and thin Marl Slate at Ferryhill is a fissile sandy limestone which has produced well-preserved fossil fishes, for example *Palaeoniscus*, and plants of a type similar to those found at the corresponding horizon in North Germany. In north-east England the Upper Permian limestones are called Magnesian Limestones. This is because they are composed principally of magnesium carbonate instead of calcium carbonate as in normal limestones. Certain of the Magnesian Limestones contain curious structures, sometimes superficially resembling fossils but in fact inorganic in origin, whilst others are associated with large-scale deposits similar in form to present-day reefs. The latter sometimes yield fairly well-preserved brachiopods, such as *Pterospirifer* and the spinose productid *Horridonia*, together with bivalves and polyzoans. The beds are exposed in the Sunderland district, as well as on the coast farther south-east. Flanking the Pennine region the normal Permian succession of marls, evaporites and, towards the top, sandstones is interrupted by coarse conglomerates, known locally as Brockrams, resulting from the erosion of adjacent land surfaces. In the Vale of Eden sandy shales, associated with these conglomerates, contain plant remains which enable the correlation of these beds with the Marl Slate on the east side of the Pennines. These beds are unfossiliferous, but an associated bed of sandy shales exposed near the village of Hilton in Westmorland is well known for its fossil plants. The underground evaporite deposits of north-east Yorkshire, comprising minerals formed by the evaporation, under arid conditions, of shallow, semi-isolated stretches of sea-water, are found in Permian rocks, whilst in parts of Scotland there occur igneous rocks of the same general age. Over most of the country, however, the Palaeozoic era closed with the formation of the New Red Sandstone, almost devoid of fossils and passing upwards without any significant break into the earliest Mesozoic strata.

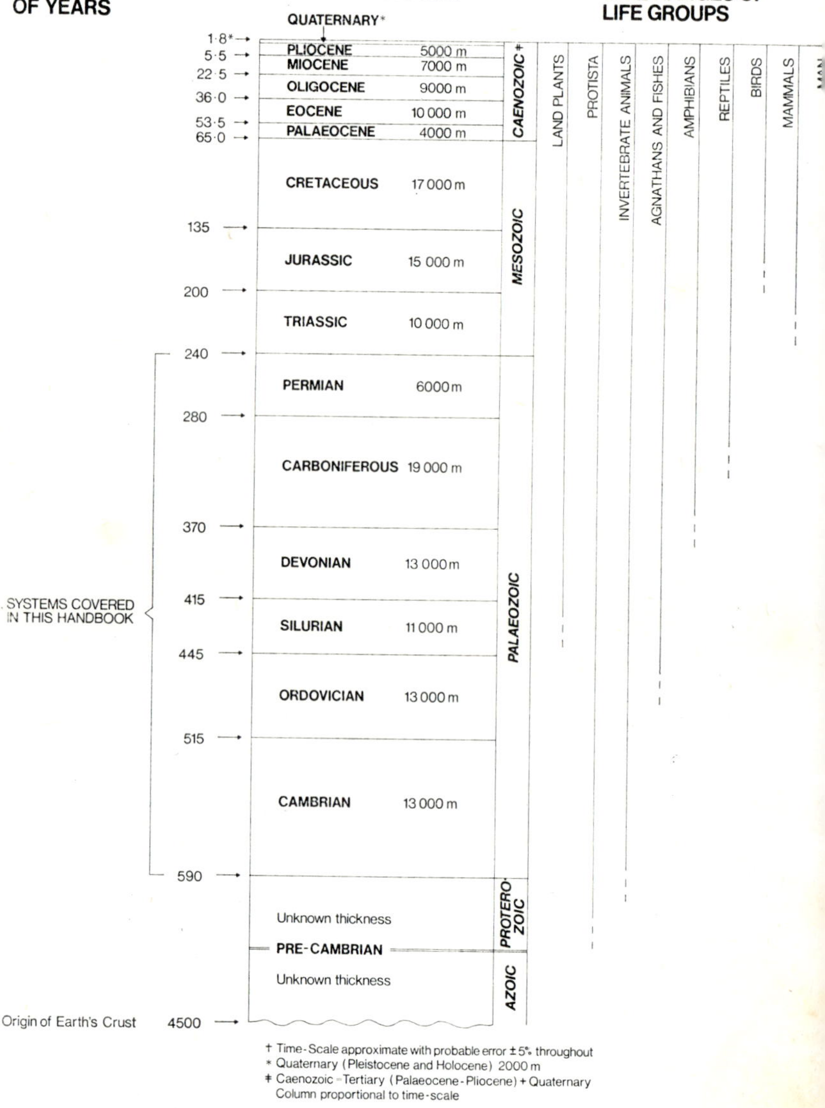

† Time-Scale approximate with probable error ± 5% throughout
* Quaternary (Pleistocene and Holocene) 2000 m
‡ Caenozoic = Tertiary (Palaeocene-Pliocene) + Quaternary
Column proportional to time-scale

Stratigraphical Tables

of

British Palaeozoic Strata

For more stratigraphical detail, the correlation charts published by the Geological Society of London, as a separate special paper for each system, should be consulted.

Cambrian System

SERIES	ZONAL GROUPS	EXAMPLES OF BRITISH STRATA
TREMADOC	*Angelina sedgwickii*	
	Shumardia pusilla	Tremadoc Slates, Shineton Shales, Upper Durness Group?
	Dictyonema flabelliforme s.l.	*Dictyonema* Beds, Manx Slates
MERIONETH	*Acerocare*	
	Peltura Zones	Dolgelly Beds, White-leaved-Oak Shales
	Leptoplastus	
	Parabolina spinulosa	*Lingula* Flags, Festiniog Beds
	Olenus	
	Agnostus pisiformis	Maentwrog Beds, Stockingford Shales
ST DAVID'S	*Paradoxides forchhammeri*	
	Paradoxides paradoxissimus	Clogau Shales, Menevian Group, Nant-pig Mudstones, Solva Group
	Paradoxides oelandicus	Barmouth Grits, Bronllwyd Grit, Upper Comley Group
COMLEY	Protolenid–Strenuellid	Rhinog Grits, Caerfai Group?, Purley Shales, Lower Comley Group
	Olenellid	Llanbedr Slates, Llanberis Slates, Lower Durness Group, Lickey Quartzite
	Tommotian Stage[1]	Dolwen Grits, Hartshill Quartzite, Hollybush Sandstone, Dalradian Group (part)

[1]The Tommotian Stage, originally defined in the U.S.S.R., is used as a term for Cambrian strata below the first occurence of trilobites

Ordovician System

SERIES	STAGES	GRAPTOLITE ZONES IN BRITAIN	EXAMPLES OF BRITISH STRATA
ASHGILL	HIRNANTIAN	*Dicellograptus anceps*	Hirnant Beds, Foel y Ddinas Mudstone Ashgill Shales
	RAWTHEYAN		Moelfryn Mudstones, Rhiwlas Limestone
	CAUTLEYAN		Slade and Redhill Beds, Drummuck Group Sholeshook Limestone, Cautley Mudstones
	PUSGILLIAN	*Dicellograptus complanatus*	Coniston Limestone, Keisley Limestone
CARADOC	ONNIAN	*Pleurograptus linearis*	Onny Shales, Nod Glas Beds Whitehouse Group, Hartfell Shales
	ACTONIAN	*Dicranograptus clingani*	Acton Scott Beds, Dufton Shales
	MARSHBROOKIAN		Cheney Longville Flags, Ardwell Group
	LONGVILLIAN		
	SOUDLEYAN	*Climacograptus wilsoni*	Horderley Sandstone, *Dicranograptus* shales
	HARNAGIAN	*Climacograptus peltifer*	Nant Hir Mudstone, Balclatchie Group Harnage Shales
	COSTONIAN	*Nemagraptus gracilis*	Coston Beds, Hoar Edge Grit Derfel Limestone
LLANDEILO			Meadowtown Beds, Hendre Shales, Glenkiln Shales, Borrowdale Volcanic Group, Mydrim Shales
		Glyptograptus teretiusculus	Llandeilo Flags
LLANVIRN		*Didymograptus murchisoni*	Ffairfach Group, Murchisoni Flags
		Didymograptus bifidus	*Bifidus* Beds
ARENIG		*Didymograptus hirundo*	Mytton Flags, *Tetragraptus* Beds Stiperstones Quartzite, Skiddaw Slate Group
		Didymograptus extensus	Henllan Ash, Upper Durness Group

Silurian System

SERIES	STAGES	GRAPTOLITE ZONES IN BRITAIN	EXAMPLES OF BRITISH STRATA
PRIDOLI (often termed Downtonian in Britain)		(no graptolites known)	Downton Group (including Ludlow Bone Bed, Downton Castle Sandstone etc.)
LUDLOW	WHITCLIFFIAN		Whitcliffe Beds, Upper Ludlow Shale Kirkby Moor Flags
		Bohemograptus bohemicus	
	LEINTWARDINIAN	*Monograptus leintwardinensis*	Leintwardine Beds
	BRINGEWOODIAN	*Monograptus scanicus*	Bringewood Beds, Aymestry Limestone
	ELTONIAN	*Pristiograptus nilssoni*	Elton Beds, Lower Ludlow Shale
WENLOCK	HOMERIAN	*Pristiograptus ludensis*	Much Wenlock Limestone
		Cyrtograptus lundgreni	Dudley Limestone
	SHEINWOODIAN	*Cyrtograptus ellesae*	
		Cyrtograptus linnarssoni	Coalbrookdale Beds, Wenlock Shale Group
		Cyrtograptus rigidus	
		Monograptus riccartonensis	
		Cyrtograptus murchisoni	Denbigh Grits, Brathay Flags
		Cyrtograptus centrifugus	Buildwas Beds, Woolhope Limestone
LLANDOVERY	TELYCHIAN	*Monoclimacis crenulata*	Purple Shale (Shropshire), May Hill Sandstone
		Monoclimacis griestoniensis	Penkill Formation
		Monograptus crispus	C_4 to C_6 beds at Llandovery, Browgill Beds
		Monograptus turriculatus	Aberystwyth Grits, *Pentamerus* Beds
	FRONIAN	*Monograptus sedgwickii*	C_1 to C_3 beds at Llandovery
	IDWIAN	*Monograptus convolutus*	B_1 to B_3 beds at Llandovery
		Monograptus gregarius	Gasworks Sandstone, Skelgill Beds
	RHUDDANIAN	*Monograptus cyphus*	Mulloch Hill Sandstone
		Monograptus vesiculosus	A_1 to A_4 beds at Llandovery
		Akidograptus acuminatus	Haverford Mudstone
		Glyptograptus persculptus	

Devonian System

STAGES	AMMONOID ZONES	BRITISH MARINE STRATA		BRITISH NON-MARINE STRATA
FAMENNIAN	*Wocklumeria*	Pilton Beds, Baggy Beds Pickwell Down Sandstone Morte Slate	OLD RED SANDSTONE	Farlow Sandstone Series
	Clymenia			Portishead Beds
	Platyclymenia			Dura Den Fish Bed
	Cheiloceras			Elgin Beds, Nairn Beds
FRASNIAN	*Manticoceras*	Ilfracombe Beds		'Orcadian Series' Caithness Flags,
GIVETIAN	*Maenioceras*	Hangman Grits		Moray Fish Beds
EIFELIAN	*Anarcestes*			Brecon Group Senni Beds, Cosheston Beds
EMSIAN	*Mimosphinctes*	Lynton Beds, Meadfoot Beds		Ridgeway Conglomerate Clee Group, Abdon Limestone
SIEGENIAN	(no ammonoids)	Dartmouth Slates		Stonehaven Beds Ditton Group, Hendre Limestone, '*Psammosteus*' Limestone
GEDINNIAN		Roseland Volcanics		St Maughan's Group

Carboniferous System

SUBSYSTEM		SUBDIVISIONS[1]	FAUNAL ZONES		EXAMPLES OF BRITISH STRATA
DINANTIAN	VISEAN	6	*Posidonia* (P) = *Dibunophyllum* (D_2)	CARBONIFEROUS LIMESTONE	Bowland Shale, Black Lias
		5	*Beyrichoceras* (B_2) = *Dibunophyllum* (D_1)		Hotwells Lst, Derbyshire 'reefs', Redesdale Lst & Ironstone, Poyllvauish Lst
		4	*Seminula* (S_2)		Cheddar Oolite, Cephalopod Shale, Great Scar Lst, Ashfell Lst
		3			
			Caninia (C_2) *Seminula* (S_1)		Burrington Oolite, Salt Hill Beds, Clitheroe 'reefs', Ashfell Lst
		2			
	TOURNAISIAN		*Caninia* (C_1)		Clifton Down Mudst, Gully Oolite, Scandel Beck Lst, Cementstones
			Zaphrentis (Z)		Black Rock Group, Colbeck Beds, Main Algal Lst, Penmaen Burrows Lst, Bewcastle Beds
		1			
			Cleistopora (K)		Lower Limestone Shales, Pinsky Gill Beds, Lynbank Beds Top Old Red Sandstone facies

[1]The subdivisions 1 to 6 represent major sedimentary cycles; A to D are arbitrary subdivisions of the Westphalian

Carboniferous System (continued)

SUBSYSTEM	SUBDIVISIONS[1]		FAUNAL ZONES	EXAMPLES OF BRITISH STRATA	
SILESIAN	STEPHANIAN			COAL MEASURES	Younger Coal Measures
	WESTPHALIAN	D			Etruria Marl Group
		C	(Many non-marine bivalve zones)		Top Marine Band Mansfield Marine Band
		B			Clay Cross Marine Band Pot Clay Marine Band
		A	*Gastrioceras* (G_2)		G Wesper Sandstone
	NAMURIAN	YEADONIAN	*Gastrioceras* (G_1)	MILLSTONE GRIT	Rough Rock
		MARSDENIAN			
		KINDERSCOUTIAN	*Reticuloceras* (R_1, R_2)		Kinderscout Grit Clayton Gill Shell Beds
		ALPORTIAN	*Homoceras* (H)		
		CHOKIERIAN			Congleton Edge Beds
		ARNSBERGIAN	*Eumorphoceras* (E_1, E_2)		Pendle Grit Grassington Grit
		PENDLEIAN			Holywell Shale, Edale Shale, Crag Lst, Great Lst, Little Lst

Permian System

DIVISIONS[1]	SUCCESSION IN Co. DURHAM[2]	OTHER EXAMPLES OF BRITISH STRATA
ZECHSTEIN	Sherwood Sandstone Upper Marls, Evaporites and Mudstone Middle Permian Marls and Upper Magnesian Limestone Hartlepool Anhydrite Middle Magnesian Limestone Lower Magnesian Limestone Marl Slate and Bressin	New Red Sandstone (lower part) St Bees Sandstone Upper Marls and Mudstone Kingscourt Gypsum Fm (Ireland) Manchester Marls St Bees Evaporites Marl Slates Brockram
ROTLIEGENDES	Yellow Sands and Breccia	Penrith Sandstone, Collyhurst Sandstone (Lowest Permian not known in Britain)

[1]Separate stages are used in the U.S.S.R., New Zealand and U.S.A., but stages are not yet internationally agreed
[2]The Durham succession is the most complete marine Permian exposed in Britain

Geological Distribution of the Species Illustrated

The numbers appended to the name of each species refer to the plate and figure which illustrate it.

CAMBRIAN SYSTEM

COMLEY SERIES

Brachiopoda	*Obolella comleyensis* Cobbold	**1,** 10, 11
Trilobita	*Olenelloides armatus* Peach	**1,** 5
	Olenellus lapworthi Peach	**1,** 6
	Protolenus latouchei Cobbold	**1,** 4
Gastropoda	*Helcionella subrugosa* (Orbigny)	**1,** 2

ST. DAVID'S SERIES

Porifera	*Protospongia fenestrata* Salter	**1,** 1
Trilobita	*Centropleura henrici* (Salter)	**1,** 3
	Eodiscus punctatus (Salter)	**2,** 4
	Paradoxides davidis Salter	**2,** 7

MERIONETH SERIES

Porifera	*Protospongia fenestrata* Salter	**1,** 1
Brachiopoda	*Lingulella davisii* (M'Coy)	**1,** 7, 8
	Orusia lenticularis (Wahlenberg)	**1,** 9
Trilobita	*Lotagnostus trisectus* (Salter)	**2,** 2
	Olenus gibbosus (Wahlenberg)	**2,** 5
	Parabolina spinulosa (Wahlenberg)	**2,** 6
	Peltura scarabaeoides (Wahlenberg)	**2,** 3
Crustacea	*Hymenocaris vermicauda* Salter	**2,** 1

TREMADOC SERIES

Hyolithida	*Hyolithes magnificus* Stubblefield & Bulman	**6,** 11
Cystoidea, Rhombifera	*Macrocystella mariae* Callaway	**3,** 6
Trilobita	*Angelina sedgwicki* Salter	**8,** 6
	Asaphellus homfrayi (Salter)	**7,** 6
	Euloma monile (Salter)	**7,** 7
	Geragnostus callavei (Lake)	**7,** 4
	Parabolinella triarthra (Callaway)	**7,** 8
	Shumardia pusilla (Sars)	**7,** 5
Crustacea	*Hymenocaris vermicauda* Salter	**2,** 1
Graptolithina	*Clonograptus tenellus* (Linnarsson)	**13,** 9
	Dictyonema flabelliforme (Eichwald)	**13,** 8

ORDOVICIAN SYSTEM

ARENIG SERIES

Brachiopoda	*Monobolina plumbea* (Salter)	**4,** 4
Trilobita	*Megalaspidella? murchisonae* (Murchison)	**8,** 7
	Neseuretus ramseyensis (Hicks)	**8,** 8
Graptolithina	*Callograptus* cf. *salteri* Hall	**13,** 1
	Dichograptus octobrachiatus (Hall)	**13,** 10
	Didymograptus hirundo (Salter)	**13,** 2
	Phyllograptus angustifolius Hall	**13,** 6
	Ptilograptus acutus (Hopkinson)	**13,** 7
	Tetragraptus serra (Brongniart)	**13,** 5

LLANVIRN SERIES

Trilobita	*Ampyx linleyensis* Whittard	**8,** 4
	Placoparia cambriensis Hicks	**8,** 5
	Placoparina shelvensis Hughes	**8,** 2, 3
	Selenopeltis inermis (Beyrich)	**9,** 8
	Stapeleyella inconstans Whittard	**8,** 1

Graptolithina	*Didymograptus bifidus* Hall	**14,** 5
	Didymograptus murchisoni (Beck)	**14,** 4

LLANDEILO SERIES

Porifera	*Ischadites koenigi* Murchison	**15,** 7
Trilobita	*Basilicus tyrannus* (Murchison)	**10,** 7
	Flexicalymene cambrensis (Salter)	**9,** 1, 2
	Marrolithus favus (Salter)	**9,** 5
	Ogygiocarella angustissima (Salter)	**10,** 6
	Platycalymene duplicata (Murchison)	**9,** 3, 4
	Selenopeltis inermis (Beyrich)	**9,** 8
Graptolithina	*Glyptograptus teretiusculus* (Hisinger)	**13,** 4
	Nemagraptus gracilis (Hall)	**14,** 6
Conodonta	*Trichonodella flexa* Rhodes	**14,** 13

CARADOC SERIES

Porifera	*Ischadites koenigi* Murchison	**15,** 7
Anthozoa	*Lyopora favosa* (M'Coy)	**3,** 4
Cystoidea, Rhombifera	*Heliocrinites* sp.	**3,** 9
Crinoidea	*Diabolocrinus globularis* (Nicholson & Etheridge)	**4,** 2
	Rhaphanocrinus basalis (M'Coy)	**4,** 3
Cyclosystoidea	*Cyclocystoides* sp.	**3,** 8
Bryozoa	*Prasopora grayae* Nicholson & Etheridge	**3,** 1–3
Brachiopoda	*Christiania perrugata* (Reed)	**5,** 1
	Dalmanella horderleyensis (Whittington)	**4,** 11–13
	Dinorthis flabellulum (J. de C. Sowerby)	**4,** 9, 10
	Harknessella vespertilio (J. de C. Sowerby)	**6,** 7–9

Brachiopoda	*Heterorthis alternata* (J. de C. Sowerby)	**4,** 6–8
	Heterorthis retrorsistria (M'Coy)	**4,** 5
	Macrocoelia expansa (J. de C. Sowerby)	**5,** 7, 8
	Nicolella actoniae (J. de C. Sowerby)	**5,** 3, 4
	Onniella broeggeri Bancroft	**5,** 2
	Reuschella horderleyensis Bancroft	**6,** 3, 4
	Sowerbyella sericea (J. de C. Sowerby)	**5,** 5, 6
	Strophomena grandis (J. de C. Sowerby)	**5,** 9, 10
Bivalvia	*Byssonychia radiata* (Hall)	**6,** 13
	Modiolopsis orbicularis (J. de C. Sowerby)	**6,** 12
Gastropoda	*Cyrtolites nodosus* (Salter)	**7,** 2
Rostroconcha	*Riberoidea lapworthi* (Etheridge)	**12,** 13
Trilobita	*Broeggerolithus broeggeri* (Bancroft)	**11,** 3–5
	Brongniartella bisulcata (M'Coy)	**11,** 6, 7
	Chasmops extensa (Boeck)	**11,** 1, 2
	Cnemidopyge bisector (Elles)	**10,** 3
	Flexicalymene caractaci (Salter)	**9,** 9
	Kloucekia apiculata (M'Coy)	**11,** 8, 9
	Ogygiocarella angustissima (Salter)	**10,** 6
	Onnia gracilis (Bancroft)	**10,** 5
	Platycalymene duplicata (Murchison)	**9,** 3, 4
	Remopleurides girvanensis Reed	**10,** 2
	Salterolithus caractaci (Murchison)	**10,** 4
	Trinucleus fimbriatus Murchison	**10,** 1
Ostracoda	*Tallinnella scripta* (Harper)	**12,** 11
Graptolithina	*Climacograptus bicornis* (Hall)	**14,** 8
	Climacograptus wilsoni Lapworth	**14,** 11
	Dicranograptus clingani Carruthers	**14,** 3
	Diplograptus multidens (Elles)	**14,** 7
	Leptograptus flaccidus (Hall)	**14,** 1
	Nemagraptus gracilis (Hall)	**14,** 6
	Orthograptus calcaratus (Lapworth)	**14,** 10
	Orthograptus truncatus (Lapworth)	**14,** 9
	Pleurograptus linearis (Carruthers)	**14,** 2

ASHGILL SERIES

Porifera	*Ischadites koenigi* Murchison	**15,** 7
Calcichordata	*Cothurnocystis elizae* Bather	**3,** 7
Cystoidea, Rhombifera	*Heliocrinites* sp.	**3,** 9
Crinoidea	*Cupulocrinus heterobrachialis* Ramsbottom	**3,** 5
	Protaxocrinus girvanensis Ramsbottom	**4,** 1
Cyclocystoidea	*Cyclocystoides* sp.	**3,** 8
Echinoidea	*Aulechinus grayae* Bather & Spencer	**3,** 10
Ophiuroidea	*Lapworthura miltoni* (Salter)	**22,** 3
Brachiopoda	*Hirnantia sagittifera* (M'Coy)	**6,** 5, 6
	Sampo ruralis (Reed)	**6,** 1, 2
Bivalvia	*Byssonychia radiata* (Hall)	**6,** 13
Gastropoda	*Cyclonema longstaffae* Lamont	**6,** 10
	Sinuites subrectangularis Reed	**7,** 1
Cephalopoda	'*Orthoceras*' *vagans* Salter	**7,** 3
Trilobita	*Corrugatagnostus sol* Whittard	**12,** 9, 10
	Cybeloides girvanensis (Reed)	**11,** 12
	Dalmanitina robertsi (Reed)	**12,** 7, 8
	Diacalymene drummuckensis (Reed)	**11,** 10, 11
	Encrinuroides sexcostatus (Salter)	**9,** 6, 7
	Flexicalymene quadrata (King)	**12,** 2
	Paraproetus girvanensis (Nicholson & Etheridge)	**12,** 1
	Phillipsinella parabola (Barrande)	**12,** 5
	Pseudosphaerexochus octolobatus (M'Coy)	**12,** 4
	Sphaerocoryphe thomsoni (Reed)	**12,** 6
	Tretaspis sortita (Reed)	**12,** 3
Ostracoda	*Primitia maccoyi* Salter	**12,** 12
Class uncertain	*Serpulites longissimus* Murchison	**17,** 1
Graptolithina	*Callograptus* cf. *salteri* Hall	**13,** 1
	Dicellograptus anceps Nicholson	**14,** 12

SILURIAN SYSTEM

LLANDOVERY SERIES

Plantae	*Mastopora fava* (Salter)	**15,** 9
Porifera	*Ischadites koenigi* Murchison	**15,** 7
Cyclocystoidea	*Cyclocystoides* sp.	**3,** 8
Ophiuroidea	*Lapworthura miltoni* (Salter)	**22,** 3
Brachiopoda	*Atrypa reticularis* (Linnaeus)	**20,** 1
	Costistricklandia lirata (J. de C. Sowerby)	**17,** 12
	Cyrtia exporrecta (Wahlenberg)	**20,** 4
	Lingula lewisi J. Sowerby	**21,** 3
	Pentamerus oblongus J. de C. Sowerby	**17,** 10, 11
	Plectatrypa imbricata (J. de C. Sowerby)	**20,** 3
	Skenidioides lewisii (Davidson)	**17,** 7
	Strophonella euglypha (Dalman)	**19,** 5, 6
	Amphistrophia aff. *funiculata* (M'Coy)	**19,** 7, 8
Bivalvia	*Pteronitella retroflexa* (Wahlenberg)	**24,** 3
Gastropoda	*Loxoplocus cancellatulus* (M'Coy)	**26,** 2, 3
Trilobita	*Calymene replicata* Shirley	**27,** 6, 7
	Dalmanites myops (König)	**28,** 5
	Encrinurus onniensis Whittard	**27,** 5
	Encrinurus punctatus (Wahlenberg)	**29,** 6
	Ktenoura retrospinosa Lane	**28,** 6
Class uncertain	*Serpulites longissimus* Murchison	**17,** 1
	Tentaculites scalaris Schlotheim	**17,** 3
Graptolithina	*Diplograptus modestus* (Lapworth)	**30,** 8
	Monograptus lobiferus (M'Coy)	**30,** 3
	Monograptus priodon (Bronn)	**30,** 2
	Monograptus sedgwickii (Portlock)	**30,** 1
	Monograptus turriculatus (Barrande)	**30,** 7
	Petalograptus minor Elles	**30,** 6
Agnatha	Thelodont scale	**37,** 1

WENLOCK SERIES

Porifera	*Ischadites koenigi* Murchison	**15,** 7
Hydrozoa	*Labechia conferta* (Lonsdale)	**16,** 10
Anthozoa	*Acervularia ananas* (Linné)	**15,** 2
	Arachnophyllum murchisoni (Edwards & Haime)	**15,** 4
	Favosites gothlandicus forma *forbesi* (Edwards & Haime)	**15,** 1
	Goniophyllum pyramidale (Hisinger)	**16,** 2
	Halysites catenularius (Linné)	**15,** 3
	Heliolites interstinctus (Linné)	**16,** 5
	Ketophyllum subturbinatum (Orbigny)	**16,** 3
	Kodonophyllum truncatum (Linné)	**16,** 7, 8
	Rhabdocyclus fletcheri (Edwards & Haime)	**16,** 4
	Syringopora bifurcata Lonsdale	**16,** 6
	Thamnopora cristata (Blumenbach)	**15,** 5
	Thecia swinderniana (Goldfuss)	**16,** 9
	Tryplasma loveni (Edwards & Haime)	**15,** 6
Calcichordata	*Placocystites forbesianus* Koninck	**23,** 4, 5
Cystoidea	*Lepocrinites quadrifasciatus* (Pearce)	**24,** 1
Crinoidea	*Crotalocrinites rugosus* (Miller)	**22,** 1, 2
	Eucalyptocrinites decorus (Phillips)	**23,** 1
	Gissocrinus goniodactylus (Phillips)	**24,** 2
	Periechocrinites moniliformis (Miller)	**23,** 2
	Sagenocrinites expansus (Phillips)	**23,** 3
Ophiuroidea	*Lapworthura miltoni* (Salter)	**22,** 3
Vermes	*Keilorites squamosus* (Phillips)	**17,** 6
	Spirorbis tenuis J. de C. Sowerby	**17,** 2
Bryozoa	*Favositella interpuncta* (Quenstedt)	**16,** 1
Brachiopoda	*Amphistrophia funiculata* (M'Coy)	**19,** 7, 8
	Anastrophia deflexa (J. de C. Sowerby)	**18,** 3
	Atrypa reticularis (Linnaeus)	**20,** 1
	Cyrtia exporrecta (Wahlenberg)	**20,** 4
	Dayia navicula (J. de C. Sowerby)	**21,** 5

Brachiopoda	*Dicoelosia biloba* (Linnaeus)	**18** 4,
	Dolerorthis rustica (J. de C. Sowerby)	**18,** 5–7
	Eoplectodonta duvalii (Davidson)	**19,** 9
	Eospirifer radiatus (J. de C. Sowerby)	**20,** 2
	Gypidula galeata (Dalman)	**18,** 2
	Howellella elegans (Muir-Wood)	**20,** 5
	Leptaena depressa (J. de C. Sowerby)	**19,** 3, 4
	Lingula lewisi J. de C. Sowerby	**21,** 3
	Meristina obtusa (J. Sowerby)	**20,** 6
	Microsphaeridiorhynchus nucula (J. de C. Sowerby)	**21,** 6, 7
	Plectatrypa imbricata (J. de C. Sowerby)	**20,** 3
	Resserella canalis (J. de C. Sowerby)	**17,** 8, 9
	Rhynchotreta cuneata (Dalman)	**18,** 1
	Skenidioides lewisii (Davidson)	**17,** 7
	Sphaerirhynchia wilsoni (J. Sowerby)	**19,** 2
	Strophonella euglypha (Dalman)	**19,** 5, 6
	Trigonirhynchia stricklandii (J. de C. Sowerby)	**19,** 1
Bivalvia	*Gotodonta ludensis* (Reed)	**24,** 7
	Grammysia cingulata (Hisinger)	**25,** 3
	Palaeopecten danbyi (M'Coy)	**25,** 4
	Pteronitella retroflexa (Wahlenberg)	**24,** 3
	Slava interrupta (Broderip)	**24,** 4
Gastropoda	'*Bembexia*' *lloydi* (J. de C. Sowerby)	**25,** 1
	Euomphalopterus alatus (Wahlenberg)	**27,** 1
	Platyceras haliotis (J. de C. Sowerby)	**26,** 6
	Poleumita discors (J. Sowerby)	**26,** 1
	Tremanotus dilatus (J. de C. Sowerby)	**25,** 2
Cephalopoda	*Dawsonoceras annulatum* (J. Sowerby)	**27,** 2
Trilobita	*Acaste downingiae* (Murchison)	**29,** 8
	Ananaspis stokesi Edwards	**29,** 4
	Bumastus barriensis (Murchison)	**28,** 7
	Calymene blumenbachi Brongniart	**28,** 1, 2
	Dalmanites myops (König)	**28,** 5
	Deiphon barrandei Whittard	**29,** 5
	Encrinurus punctatus (Wahlenberg)	**29,** 6
	Encrinurus variolaris (Brongniart)	**29,** 7

Trilobita	*Ktenoura retrospinosa* Lane	**28,** 6
	Leonaspis deflexa (Lake)	**29,** 4
	Sphaerexochus britannicus Dean	**28,** 4
	Trimerus delphinocephalus (Green)	**28,** 3
Ostracoda	*Beyrichia* cf. *kloedeni* M'Coy	**29,** 1, 2
	Leperditia balthica (Hisinger)	**29,** 3
Class uncertain	*Cornulites serpularius* Schlotheim	**17,** 5
	Serpulites longissimus Murchison	**17,** 1
	Tentaculites ornatus J. de C. Sowerby	**17,** 4
	Tentaculites scalaris Schlotheim	**17,** 3
Graptolithina	*Cyrtograptus murchisoni* Carruthers	**30,** 9
	Monograptus priodon (Bronn)	**30,** 2
	Monograptus turriculatus (Barrande)	**30,** 7
Agnatha	Thelodont scale	**37,** 1

LUDLOW SERIES

Porifera	*Amphispongia oblonga* Salter	**15,** 8
	Ischadites koenigi Murchison	**15,** 7
Anthozoa	*Favosites gothlandicus* forma *forbesi* (Edwards & Haime)	**15,** 1
Echinoidea	*Palaeodiscus ferox* Salter	**21,** 1
Ophiuroidea	*Lapworthura miltoni* (Salter)	**22,** 3
Vermes	*Keilorites squamosus* (Phillips)	**17,** 6
Brachiopoda	*Atrypa reticularis* (Linnaeus)	**20,** 1
	Cyrtia exporrecta (Wahlenberg)	**20,** 4
	Dayia navicula (J. de C. Sowerby)	**21,** 5
	Dicoelosia biloba (Linnaeus)	**18,** 4
	Eospirifer radiatus (J. de C. Sowerby)	**20,** 2
	Howellella elegans (Muir-Wood)	**20,** 5
	Kirkidium knighti (J. Sowerby)	**21,** 11, 12
	Leptaena depressa (J. de C. Sowerby)	**19,** 3, 4
	Lingula lewisi J. de C. Sowerby	**21,** 3
	Meristina obtusa (J. Sowerby)	**20,** 6

Group	Species	Plate, figs
Brachiopoda	*Microsphaeridiorhynchus nucula* (J. de C. Sowerby)	**21,** 6, 7
	Protochonetes ludloviensis Muir-Wood	**21,** 4
	Resserella canalis (J. de C. Sowerby)	**17,** 8, 9
	Salopina lunata (J. de C. Sowerby)	**21,** 8–10
	Shaleria ornatella (Davidson)	**21,** 2
	Sphaerirhynchia wilsoni (J. Sowerby)	**19,** 2
	Strophonella euglypha (Dalman)	**19,** 5, 6
Bivalvia	*Fuchsella amygdalina* (J. de C. Sowerby)	**24,** 6
	Goniophora cymbaeformis (J. de C. Sowerby)	**24,** 5
	Gotodonta ludensis (Reed)	**24,** 7
	Grammysia cingulata (Hisinger)	**25,** 3
	Palaeopecten danbyi (M'Coy)	**25,** 4
	Pteronitella retroflexa (Wahlenberg)	**24,** 3
	Slava interrupta (Broderip)	**24,** 4
Gastropoda	'*Bembexia*' *lloydi* (J. de C. Sowerby)	**25,** 1
	Euomphalopterus alatus (Wahlenberg)	**27,** 1
	Loxonema gregaria (J. de C. Sowerby)	**26,** 4
	Platyceras haliotis (J. de C. Sowerby)	**26,** 5
	Poleumita discors (J. Sowerby)	**26,** 1
	Tremanotus dilatus (J. de C. Sowerby)	**25,** 2
Cephalopoda	*Dawsonoceras annulatum* (J. Sowerby)	**27,** 2
	Gomphoceras ellipticum M'Coy	**27,** 3
Trilobita	*Dalmanites myops* (König)	**28,** 5
	Delops obtusicaudatus (Salter)	**29,** 9, 10
Ostracoda	*Leperditia balthica* (Hisinger)	**29,** 3
Other Arthropoda	*Ceratiocaris stygia* Salter	**30,** 13
	Erretopterus bilobus Salter	**30,** 14
Class uncertain	*Cornulites serpularius* Schlotheim	**17,** 5
	Serpulites longissimus Murchison	**17,** 1
	Tentaculites scalaris Schlotheim	**17,** 3
Graptolithina	*Monograptus colonus* (Barrande)	**30,** 4
	Monograptus leintwardinensis Hopkinson	**30,** 5

Conodonta	*Ozarkodina typica* Branson & Mehl	**30,** 11
	Panderodus unicostatus (Branson & Mehl)	**30,** 10
	Spathognathodus typicus (Branson & Mehl)	**30,** 12
Agnatha	Thelodont scale	**37, 1**

DEVONIAN SYSTEM

(a) MARINE DEVONIAN

Anthozoa	*Digonophyllum bilaterale* (Champernowne)	**32,** 6
	Disphyllum goldfussi (Geinitz)	**32,** 5
	Favosites goldfussi Orbigny	**31,** 5
	Heliolites porosus (Goldfuss)	**31,** 6
	Hexagonaria goldfussi (de Verneuil & Haime)	**31,** 7
	Pachyphyllum devoniense Edwards & Haime	**32,** 7
	Phillipsastraea hennahi (Lonsdale)	**31,** 9
Hydrozoa	*Stromatopora huepschii* (Bargatsky)	**31,** 4
	Thamnopora cervicornis (Blainville)	**31,** 8
Crinoidea	*Hexacrinites interscapularis* (Phillips)	**32,** 8
Brachiopoda	*Camarotoechia laticosta* (Phillips)	**34,** 7
	Cyrtina heteroclita Defrance	**33,** 2
	Cyrtospirifer extensus (J. de C. Sowerby)	**33,** 1
	Hypothyridina cuboides (J. de C. Sowerby)	**34,** 5
	Ladogia triloba (J. de C. Sowerby)	**34,** 6
	Mesoplica praelonga (J. de C. Sowerby)	**32,** 2
	Plectatrypa aspera (Schlotheim)	**34,** 4
	Pyramidalia simplex (Phillips)	**33,** 3
	Productella fragaria (J. de C. Sowerby)	**32,** 1
	Rhenorensselaeria strigiceps (Roemer)	**33,** 8
	Sieberella brevirostris (Phillips)	**33,** 5, 6
	Spirifer undiferus Roemer	**33,** 4

Brachiopoda	*Stringocephalus burtini* Defrance	**34,** 3
	Stropheodonta nobilis (M'Coy)	**33,** 7
	Uncites gryphus Schlotheim	**34,** 1, 2
Bivalvia	*Actinopteria placida* (Whidborne)	**35,** 3
	Buchiola retrostriata (Buch)	**35,** 2
	'*Cucullaea*' *unilateralis* J. de C. Sowerby	**35,** 1
Gastropoda	*Euryzone delphinuloides* (Schlotheim)	**35,** 6
	Murchisonia bilineata (Dechen)	**35,** 5
	Serpulospira militaris (Whidborne)	**35,** 4
Cephalopoda	*Manticoceras intumescens* (Beyrich)	**35,** 7
	Tornoceras psittacinum (Whidborne)	**36,** 9
Trilobita	*Crotalocephalus pengellii* (Whidborne)	**36,** 7
	Dechenella setosa (Whidborne)	**36,** 4, 5
	Phacops accipitrinus (Phillips)	**36,** 6
	Scutellum costatum (Goldfuss)	**36,** 3
	Trimerocephalus mastophthalmus (Richter)	**36,** 1, 2
Conodonta	*Icriodus* sp.	**36,** 8

(b) OLD RED SANDSTONE

Plantae	*Psilophyton princeps* Dawson	**31,** 1
	Zosterophyllum llanoveranum Croft & Lang	**31,** 2, 3
Brachiopoda	*Lingula cornea* J. de C. Sowerby	**32,** 3
	Lingula minima J. de C. Sowerby	**32,** 4
Bivalvia	*Archanodon jukesi* (Baily)	**35,** 8
Agnatha	*Pteraspis rostrata* (Agassiz) *trimpleyensis* White	**37,** 3
	Cephalaspis lyelli Agassiz	**37,** 2
	Thelodont scale	**37,** 1
Pisces	*Asterolepis maxima* (Agassiz)	**37,** 5
	Coccosteus cuspidatus Miller	**36,** 10
	Holoptychius giganteus Agassiz	**37,** 4

CARBONIFEROUS SYSTEM

LOWER CARBONIFEROUS

Plantae	*Rhacopteris petiolata* Goeppert	**41,** 2
	Rhodea tenuis Gothan	**40,** 4
	Telangium affine (Lindley & Hutton)	**39,** 1
Foraminifera	*Archaediscus karreri* Brady	**42,** 1, 2
	Climacammina antiqua Brady	**42,** 12, 13
	Endothyranopsis crassa (Brady)	**42,** 5
	Howchinia bradyana (Howchin)	**42,** 10, 11
	Lugtonia concinna (Brady)	**42,** 14
	Plectogyra bradyi (Mikhailov)	**42,** 6
	Stacheia pupoides Brady	**42,** 8, 9
	Stacheoides polytremoides (Brady)	**42,** 7
	Tetrataxis conica Ehrenberg	**42,** 3, 4
Porifera	*Hyalostelia smithi* Young & Young	**41,** 1
Anthozoa	*Amplexizaphrentis enniskilleni* (Edwards & Haime) var. *derbiensis* Lewis	**44,** 3
	Amplexus coralloides J. Sowerby	**44,** 1
	Aulophyllum fungites (Fleming)	**44,** 2
	Caninia cylindrica (Scovler)	**45,** 2
	Dibunophyllum bipartitum (M'Coy)	**43,** 1
	Lithostrotion junceum (Fleming)	**43,** 2, 3
	Lithostrotion portlocki (Bronn)	**43,** 4, 5
	Lonsdaleia floriformis (Fleming)	**44,** 4
	Michelinia tenuisepta (Phillips)	**45,** 4
	Palaeosmilia murchisoni Edwards & Haime	**44,** 6, 7
	Palaeosmilia regium (Phillips)	**44,** 5
	Syringopora geniculata Phillips	**45,** 3
Blastoidea	*Codaster acutus* M'Coy	**59,** 3, 4
	Orbitremites ellipticus (G. B. Sowerby)	**59,** 5, 6
	Orophocrinus verus (Cumberland)	**59,** 7
Crinoidea	*Actinocrinites triacontadactylus* Miller	**61,** 1
	Amphoracrinus gigas Wright	**60,** 3
	Gilbertsocrinus konincki Grenfell	**61,** 2
	Platycrinites gigas Phillips	**60,** 4

Conulata	*Conularia quadrisulcata* (J. de C. Sowerby)	**45,** 5
Echinoidea	*Archaeocidaris urii* (Fleming)	**59,** 9
	Archaeocidaris sp.	**59,** 8
	Lovenechinus lacazei (Julien)	**59,** 10
	Melonechinus etheridgei (Keeping)	**60,** 1
Vermes	*Spirobis pusillus* (Martin)	**46,** 11
Bryozoa	*Fenestella plebeia* M'Coy	**45,** 1
Brachiopoda	*Actinoconchus lamellosus* (Léviellé)	**52,** 4
	Antiquatonia hindi (Muir-Wood)	**47,** 4
	Brachythyris pinguis (J. Sowerby)	**49,** 3
	Composita ambigua (J. Sowerby)	**50,** 1
	Daviesiella llangollensis (Davidson)	**49,** 4
	Dictyoclostus semireticulatus (Martin)	**47,** 5
	Dielasma hastatum (J. de C. Sowerby)	**52,** 5
	Eomarginifera setosa (Phillips)	**46,** 5, 6
	Gigantoproductus giganteus (J. Sowerby)	**47,** 6
	Krotovia spinulosa (J. Sowerby)	**47,** 3
	Leptagonia analoga (Phillips)	**48,** 1, 2
	Lingula mytilioides J. Sowerby	**46,** 8
	Lingula squamiformis Phillips	**46,** 9
	Linoprotonia corrugatus (M'Coy)	**46,** 10
	Martinia glabra (Martin)	**50,** 2
	Orbiculoidea nitida (Phillips)	**46,** 7
	Overtonia fimbriata (J. de C. Sowerby)	**46,** 1
	Phricodothyris lineata (J. Sowerby)	**48,** 7, 8
	Productus productus (Martin)	**46,** 2–4
	Pugnax acuminatus (J. Sowerby)	**52,** 3
	Pleuropugnoides pleurodon (Phillips)	**51,** 2
	Punctospirifer scabricostus North	**52,** 1
	Pustula pustulosa (Phillips)	**48,** 3, 4
	Rhipidomella michelini (Léveillé)	**48,** 5, 6
	Rugosochonetes hardrensis (Phillips)	**47,** 1
	Schellwienella crenistria (Phillips)	**49,** 2
	Schizophoria resupinata (Martin)	**49,** 1
	Spirifer attenuatus J. de C. Sowerby	**52,** 2
	Spirifer striatus (Martin)	**50,** 3
	Syringothyris cuspidata (J. Sowerby)	**51,** 3

Bivalvia	*Aviculopecten plicatus* (J. de C. Sowerby)	**54,** 7
	Edmondia sulcata (Phillips)	**54,** 2
	Lithophaga lingualis (Phillips)	**53,** 2
	Polidevcia attenuata (Fleming)	**53,** 1
	Posidonia becheri Bronn	**53,** 3
	Posidoniella vetusta (J. de C. Sowerby)	**53,** 4
	Pterinopectinella granosa (J. de C. Sowerby)	**54,** 5
	Sanguinolites costellatus M'Coy	**54,** 4
	Wilkingia elliptica (Phillips)	**54,** 3
Rostroconcha	*Conocardium hibernicum* J. Sowerby	**53,** 5, 6
Gastropoda	*Euconospira conica* (Phillips)	**56,** 2
	Glabrocingulum armstrongi Thomas	**56,** 9
	Glabrocingulum atomarium (Phillips)	**56,** 4
	Mourlonia carinata (J. Sowerby)	**56,** 7
	Naticopsis elliptica (Phillips)	**57,** 1
	Palaeostylus rugiferus (Phillips)	**57,** 2
	Platyceras vetustum (J. de C. Sowerby)	**56,** 5
	Soleniscus acutus (J. de C. Sowerby)	**56,** 6
	Straparollus dionysii de Montfort	**56,** 3
	Straparollus pentangulatus (J. Sowery)	**56,** 8
Monoplacophora	*Euphenites urei* (Fleming)	**56,** 1
Cephalopoda	*Beyrichoceras obtusum* (Phillips)	**58,** 4, 5
	Goniatites crenistria Phillips	**59,** 1
	Muensteroceras truncatum (Phillips)	**58,** 2, 3
	Neoglyphioceras spirale (Phillips)	**59,** 2
Trilobita	*Brachymetopus ouralicus* (Verneuil) *ornatus* Woodward	**62,** 3, 4
	Cummingella jonesi (Portlock)	**62,** 2
	Eocyphinium seminiferum (Phillips)	**62,** 1
	Phillipsia gemmulifera (Phillips)	**62,** 5
	Cummingella jonesi (Portlock) *laticaudata* (Woodward)	**62,** 8, 9
	Spatulina spatulata (Woodward)	**62,** 6, 7
Ostracoda	*Entomoconchus scouleri* M'Coy	**63,** 1
	Richteria biconcentrica (Jones)	**63,** 2
Other Crustacea	*Perimecturus parki* (Peach)	**62,** 10

Pisces	*Cladodus mirabilis* Agassiz	**63,** 9
	Gyracanthus formosus Agassiz	**64,** 5
	Helodus turgidus (Agassiz)	**64,** 3
	Orodus ramosus Agassiz	**64,** 4
	Psammodus rugosus Agassiz	**64,** 2
	Psephodus magnus (Portlock)	**64,** 1
	Rhizodus hibberti (Agassiz)	**65,** 4

UPPER CARBONIFEROUS

Plantae	*Alethopteris serli* Brongniart	**40,** 5
	Annularia stellata (Schlotheim)	**38,** 1
	Asterophyllites equisetiformis (Schlotheim)	**38,** 4
	Calamites suckowi Brongniart	**38,** 5
	Cyclopteris trichomanoides Brongniart	**41,** 4
	Cordaites angulosostriatus Grand' Eury	**41,** 5
	Lepidodendron aculeatum Sternberg	**39,** 5
	Lepidodendron sternbergi Brongniart	**39,** 6
	Mariopteris nervosa (Brongniart)	**40,** 2, 3
	Neuropteris gigantea Sternberg	**41,** 3
	Pecopteris polymorpha Brongniart	**40,** 1
	Sigillaria mamillaris Brongniart	**39,** 4
	Sphenophyllum emarginatum Brongniart	**38,** 2
	Sphenopteris alata Brongniart	**39,** 2
	Stigmaria ficoides Brongniart	**38,** 3
	Trigonocarpus sp.	**39,** 3
Foraminifera	*Lugtonia concinna* (Brady)	**42,** 14
	Stacheioides pupoides Brady	**42,** 8, 9
	Stacheoides polytremoides (Brady)	**42,** 7
	Tetrataxis conica Ehrenberg	**42,** 3, 4
Anthozoa	*Dibunophyllum bipartitum* (M'Coy)	**43,** 1
	Lithostrotion junceum (Fleming)	**43,** 2, 3
	Lonsdaleia floriformis (Fleming)	**44,** 4
	Palaeosmilia regium (Phillips)	**44,** 5
Brachiopoda	*Brachythyris pennystonensis* (George)	**51,** 1
	Composita ambigua (J. Sowerby)	**50,** 1
	Dictyoclostus semireticulatus (Martin)	**47,** 5
	Dielasma hastatum (J. de C. Sowerby)	**52,** 5

Brachiopoda	*Eomarginifera setosa* (Phillips)	**46,** 5, 6
	Krotovia spinulosa (J. Sowerby)	**47,** 3
	Leptagonia analoga (Phillips)	**48,** 1, 2
	Lingula mytiloides J. Sowerby	**46,** 8
	Lingula squamiformis Phillips	**46,** 9
	Martinia glabra (Martin)	**50,** 2
	Orbiculoidea nitida (Phillips)	**46,** 7
	Overtonia fimbriata (J. de C. Sowerby)	**46,** 1
	Phricodothyris lineata (J. Sowerby)	**48,** 7, 8
	'*Productus*' *craigmarkensis* (Muir-Wood)	**47,** 2
	Productus productus (Martin)	**46,** 2–4
	Pugnoides pleurodon (Phillips)	**51,** 2
	Rhipidomella michelini (Léveillé)	**48,** 5–6
	Schellwienella crenistria (Phillips)	**49,** 2
	Schizophoria resupinata (Martin)	**49,** 1
Bivalvia	*Anthraconaia adamsi* (Salter)	**55,** 7
	Anthracosia atra (Trueman)	**55,** 3
	Anthracosia planitumida (Trueman)	**55,** 4
	Anthracosphaerium exiguum (Davies & Trueman)	**55,** 2
	Carbonicola communis Davies & Trueman	**55,** 5
	Carbonicola pseudorobusta Trueman	**55,** 6
	Dunbarella papyracea (J. de C. Sowerby)	**54,** 6
	Naiadites modiolaris J. de C. Sowerby	**55,** 1
	Polidevcia attenuata (Fleming)	**53,** 1
	Sanguinolites costellatus M'Coy	**54,** 4
	Schizodus carbonarius (J. de C. Sowerby)	**54,** 1
	Wilkingia elliptica (Phillips)	**54,** 3
Gastropoda	*Glabrocingulum armstrongi* Thomas	**56,** 9, 10
	Palaeostylus rugiferus (Phillips)	**57,** 2
Cephalopoda	*Gastrioceras carbonarium* (Buch)	**58,** 6
	Homoceras diadema (Beyrich)	**58,** 1
	Reticuloceras bilingue (Salter)	**57,** 4
	Reticuloceras reticulatum (Phillips)	**57,** 3

Arthropoda	*Acantherpestes ferox* (Salter)	**63,** 5
	Ectodemites bipartitus (Vine)	**63,** 3
	Eophrynus prestvici (Buckland)	**63,** 6
	Euproops rotundatus (Prestwich)	**63,** 4
Conulata	*Conularia quadrisulcata* J. Sowerby	**45,** 5
Conodonta	*Gnathodus bilineatus* (Roundy)	**63,** 8
	Idiognathoides corrugata (Harris & Hollingsworth)	**63,** 7
Pisces	*Gyracanthus formosus* Agassiz	**64,** 5
	Megalichthys hibberti Agassiz	**65,** 1, 2
	Xenacanthus laevissimus (Agassiz)	**64,** 6
	Rhabdoderma tingleyense (Davis)	**65,** 3
	Sagenodus inaequalis Owen	**66,** 1
Amphibia	Anthracosaurian vertebra	**66,** 2
	Keraterpeton galvani Huxley	**66,** 3
	Megalocephalus cf. *macromma* Barkas	**66,** 4

PERMIAN SYSTEM

Foraminifera	*Nodosinella digitata* Brady	**67,** 5, 6
Bryozoa	*Fenestella retiformis* (Schlotheim)	**67,** 7, 8
Brachiopoda	*Dielasma elongatum* (Schlotheim)	**68,** 2
	Horridonia horrida (J. Sowerby)	**68,** 4, 5
	Orthothrix excavata (Geinitz)	**67,** 1, 2
	Pterospirifer alatus (Schlotheim)	**68,** 1
	Spiriferellina cristata (Schlotheim)	**68,** 3
	Stenoscisma humbletonensis (Howse)	**67,** 3, 4
Bivalvia	*Bakevellia binneyi* (Brown)	**69,** 3, 4
	Parallelodon striatus (Schlotheim)	**69,** 1
	Permophorus costatus (Brown)	**69,** 2
	Pseudomonotis speluncularia (Schlotheim)	**69,** 6
	Schizodus obscurus (J. Sowerby)	**69,** 5
Pisces	*Palaeoniscus freieslebenensis* Blainville	**69,** 7

The Scientific Names of Fossils

The scientific name of a species is established by its publication with a description of the distinctive characters and preferably also with an illustration of the species. The worker describing the latter is alluded to as its author. The name of each species consists essentially of two words which are either Latin or treated as Latin. The first word is the name of the genus to which the species is assigned, and the second (the specific name) denotes the species. Sometimes the species of a genus are grouped in subgenera which also have Latin names. The subgeneric name is then placed between the generic and specific names, but in round brackets. The name of the author of a species is usually placed after the specific name; this gives a clue to where the description of a species is to be found. If the species has been transferred to a different genus from that under which it was originally described, its author's name is placed in round brackets. If the specific name is an adjective, it must agree in gender with the generic name. Some specific names, however, are nouns in apposition to the generic name and are not liable to change according to the gender of the latter.

Sometimes it is desired to indicate that a specimen belongs to a definite subspecies, that is, a group in which, perhaps, geographical isolation or evolutionary changes have resulted in slight differences from typical specimens of the species. In such cases a Latin subspecific name is used, and this and its author's name follow the names already mentioned.

The same genus or species has sometimes been described by different workers under different names, and in such cases, except in certain circumstances, the name first used must be accepted. The discovery of earlier names has thus been one reason for changes in nomenclature. A more important cause of changes in the name of organisms lies in the fact that nomenclature is dependent upon classification. Increased knowledge of a species may show that it was referred by its author to a genus with which it has no affinity, as in the case of the species from the Coal Measures which was originally called *Unio acutus* but is now known not to belong even to the same family as *Unio* and is placed in the genus *Carbonicola*. Moreover, one worker will treat a group of species as a distinct genus, whereas another will include the same group in a genus described earlier. Similarly, one worker will unite in a single species a series of specimens which another will consider to belong to two or more distinct species.

While many goniatites were formerly included in a single genus *Goniatites* and many Palaeozoic brachiopods in *Dalmanella*, *Orthis*, *Productus* or *Spirifer*, species of these groups are now classified in a great number of different genera. The views of modern authorities have been the main criterion in deciding what names should be used for any species illustrated in the present handbook. Other names which have been used from time to time (and which may differ in either the generic or the specific name, or in both) are, if thought important enough, cited as synonyms (abbreviation 'syn'.).

L.R.C.

Explanation of Plates

The geological range given for each species is that at present known and applies only to Great Britain.

Two or more drawings bearing the same number and linked by a broken line are views of the same specimen.

The names quoted in square brackets after the abbreviation Syn. (=Synonym) are other names that have been used, sometimes incorrectly, for the species (see p. 48).

An asterisk by the name of a species indicates that it may be found exhibited near the entrance to the Fossil Mammal Gallery at the British Museum (Natural History).

E

Plate 1

Cambrian Sponge (Fig. 1), Monoplacophoran (Fig. 2), Trilobites (Figs. 3–6) and Brachiopods (Figs. 7–11)

1.* **Protospongia fenestrata** Salter. Group of spicules (×2½.) Middle Cambrian; St David's, Dyfed. RANGE: Middle-Upper Cambrian.

2. **Helcionella subrugosa** (d'Orbigny). (×1½.) Lower Cambrian; near Church Stretton, Salop. RANGE: Lower Cambrian. [Syn., *Helcion rugosa* (Hall).]

3. **Centropleura henrici** (Salter). Cranidium (×¾.) Middle Cambrian; St David's, Dyfed. RANGE: Middle Cambrian. [Syn., *Anopolenus henrici*.]

4. **Protolenus latouchei** Cobbold. Cephalon (×2.) Lower Cambrian; Comley, near Church Stretton, Salop. RANGE: Lower Cambrian.

5.* **Olenelloides armatus** Peach. (×1½.) Lower Cambrian; Meall a' Ghiubhais, near Kinlochewe, Ross-shire. RANGE: Lower Cambrian.

6.* **Olenellus lapworthi** Peach. (×1½.) Lower Cambrian; Meall a' Ghiubhais, near Kinlochewe, Ross-shire. RANGE: Lower Cambrian.

7, 8.* **Lingulella davisii** (M'Coy). Upper Cambrian. 7 (×2.) Tremadoc, Gwynedd. 8, interior of ventral valve (×1½.) Dolgelly, Gwynedd. RANGE: Genus, Upper Cambrian–Ordovician; Species, Upper Cambrian. [Syn., *Lingula davisii*].

9. **Orusia lenticularis** (Wahlenberg). Dorsal valve (×4.) Upper Cambrian; near Tremadoc, Gwynedd. RANGE: Upper Cambrian.

10, 11.* **Obolella comleyensis** Cobbold. 10, ventral valve (×5.) 11, internal mould of dorsal valve (×8.) Lower Cambrian; Comley, near Church Stretton, Salop. RANGE: Lower Cambrian.

Plate 1

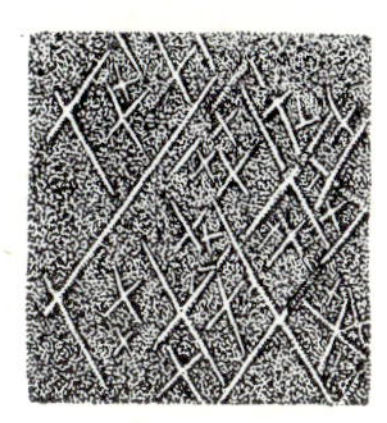

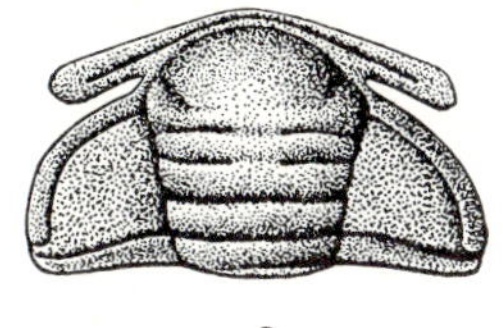

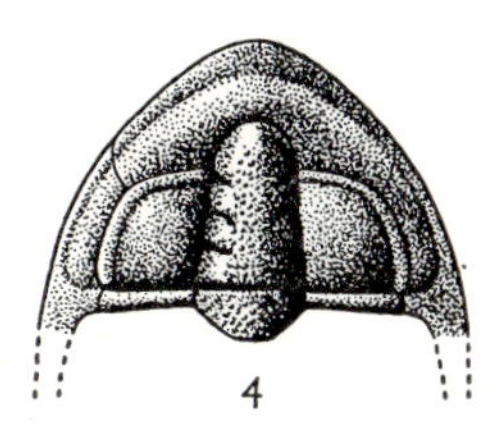

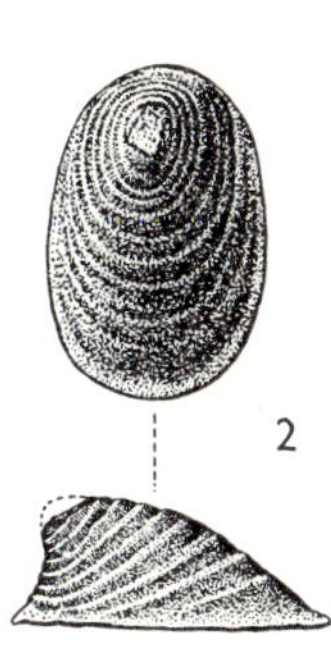

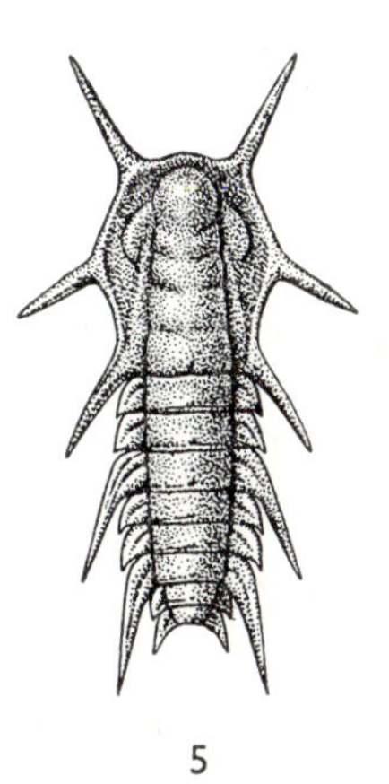

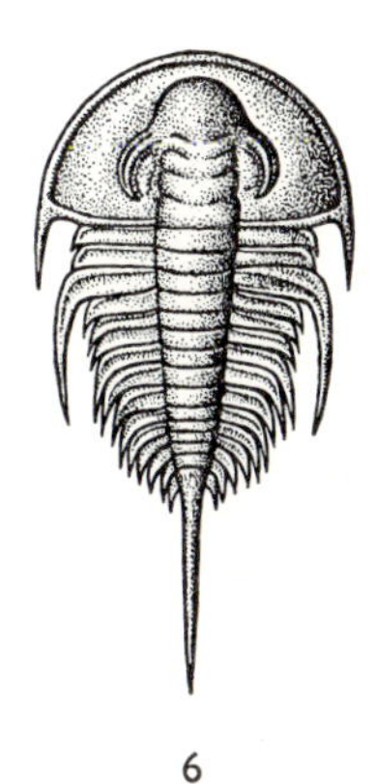

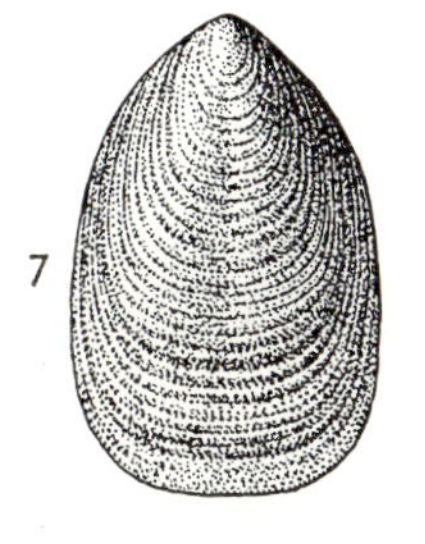

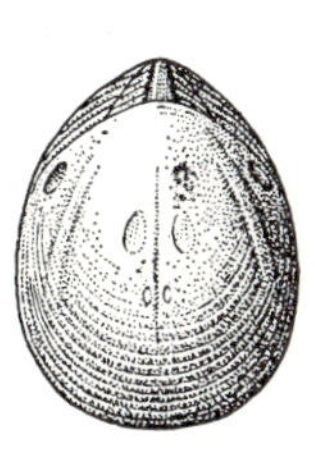

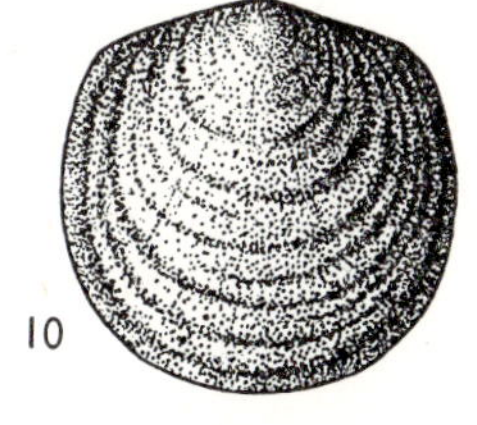

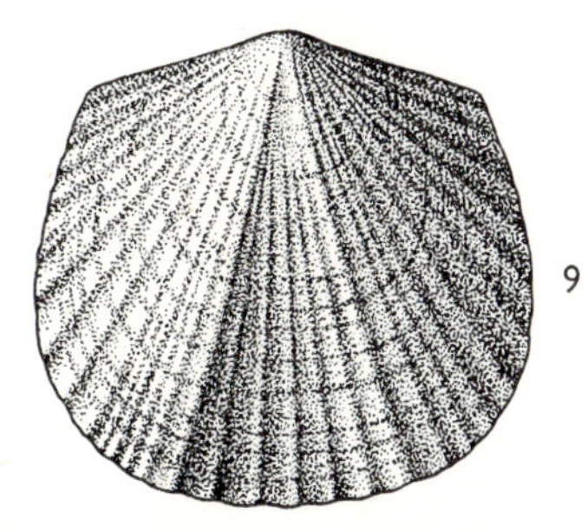

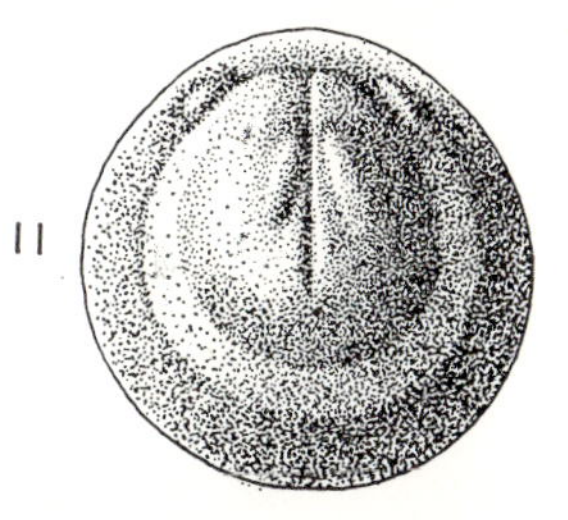

Plate 2

Cambrian Malacostracan Crustacean (Fig. 1) and Trilobites (Figs. 2–7)

1. **Hymenocaris vermicauda** Salter. (×1.) Upper Cambrian; Borth-y-Gest, Portmadoc, Gwynedd. RANGE: Merioneth and Tremadoc Series.

2. **Lotagnostus trisectus** (Salter). (×3.) Upper Cambrian; south-western end of Malvern Hills. RANGE: Upper Cambrian., [Syn., *Agnostus trisectus*.]

3.* **Peltura scarabaeoides** (Wahlenberg). (×4.) Upper Cambrian; Dolgelly, Gwynedd. RANGE: Upper Cambrian.

4. **Eodiscus punctatus** (Salter). (×3.) Middle Cambrian; St David's, Dyfed. RANGE: Middle Cambrian. [Syn., *Microdiscus punctatus*.]

5. **Olenus gibbosus** (Wahlenberg). (×1¼.) Upper Cambrian; Dolgelly, Gwynedd. RANGE: Upper Cambrian.

6.* **Parabolina spinulosa** (Wahlenberg). (×2.) Upper Cambrian; Dolgelly, Gwynedd. RANGE: Upper Cambrian.

7.* **Paradoxides davidis** Salter. (×½.) Middle Cambrian; St David's, Dyfed. RANGE: Middle Cambrian.

Plate 2

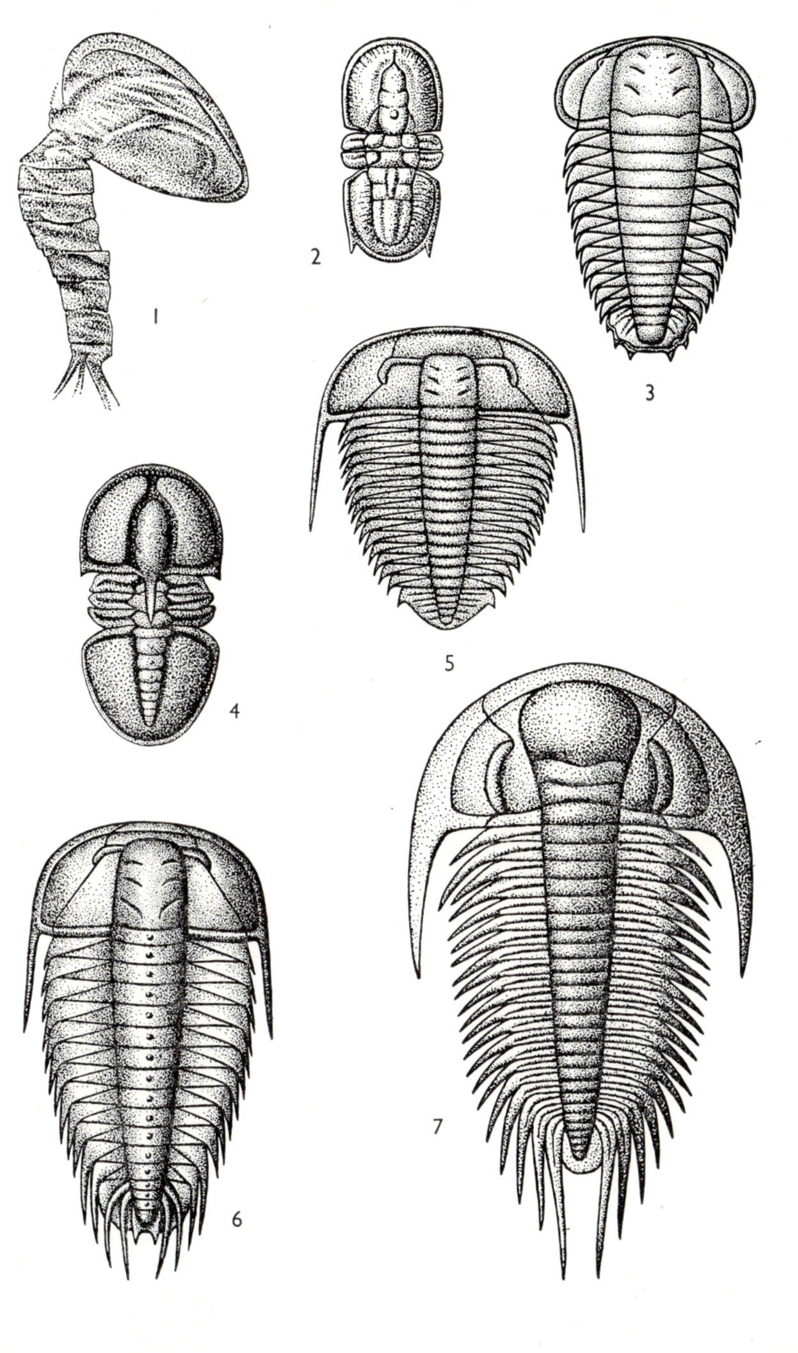

Plate 3

Cambrian Echinoderm (Fig. 6) and Ordovician Bryozoan (Figs. 1–3), Coral (Fig. 4), Echinoderms (Figs. 5, 8–10) and Calcichordate (Fig. 7)

1–3.* **Prasopora grayae** Nicholson & Etheridge. 1 (×1.) 2,3, transverse, and longitudinal sections (×15.) Caradoc Series; Craighead, Girvan, Ayrshire. RANGE: Caradoc Series.

4.* **Lyopora favosa** (M'Coy). (×3½.) Caradoc Series; Craighead, Girvan, Ayrshire. RANGE: Caradoc Series.

5.* **Cupulocrinus heterobrachialis** Ramsbottom. Artificial cast (×1.) Ashgill Series; Thraive Glen, Girvan, Ayrshire. RANGE: Ashgill Series.

6. **Macrocystella mariae** Callaway. Artificial cast (×1.) Tremadoc Series; Sheinton, Salop. RANGE: Tremadoc Series.

7.* **Cothurnocystis elizae** Bather. (×1½.) Ashgill Series; Thraive Glen, Girvan, Ayrshire. RANGE: Ashgill Series.

8. **Cyclocystoides** sp. Artificial cast. (×3½.) Ashgill Series; Thraive Glen, Girvan, Ayrshire. RANGE: Caradoc Series–Silurian, Llandovery Series.

9.* **Heliocrinites** sp. (×1.) Probably Ashgill Series; Nant Fawr Waterfall, Bwlch-y-Gaseg, near Cynwyd, Gwynedd. RANGE: Caradoc–Ashgill Series.

10.* **Aulechinus grayae** Bather & Spencer. (×1¼.). Ashgill Series; Thraive Glen, Girvan, Ayrshire. RANGE: Ashgill Series.

Plate 3

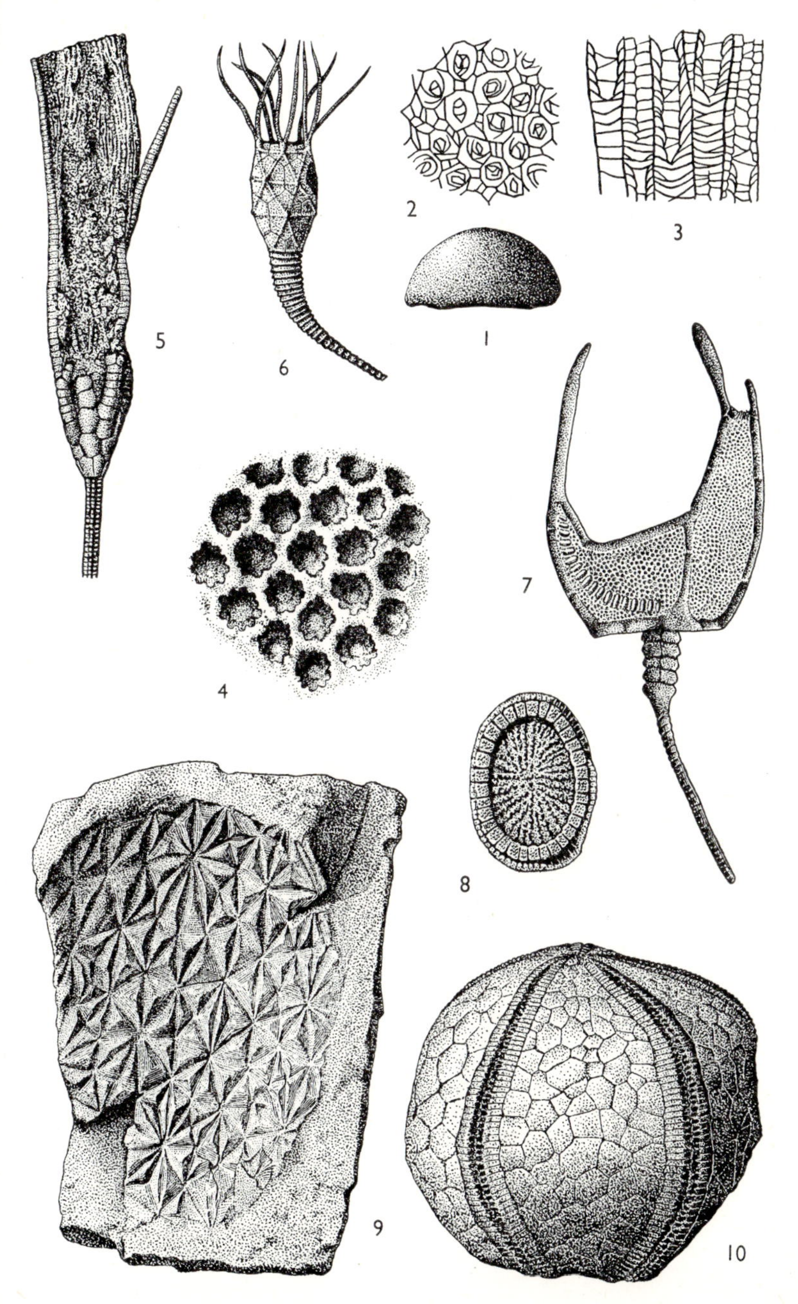

Plate 4

Ordovician Crinoids (Figs. 1–3) and Brachiopods (Figs. 4–13)

1.* **Protaxocrinus girvanensis** Ramsbottom. Artificial cast (×1½.) Ashgill Series; Thraive Glen, Girvan, Ayrshire. RANGE: Ashgill Series.

2. **Diabolocrinus globularis** (Nicholson & Etheridge). (×2.) Caradoc Series; Craighead, Girvan, Ayrshire. RANGE: Caradoc Series.

3. **Rhaphanocrinus basalis** (M'Coy). (×1.) Caradoc Series; near Church Stretton, Salop. RANGE: Caradoc Series. [Syn., *Glypto-crinus basalis, Balacrinus basalis.*]

4.* **Monobolina plumbea** (Salter). (×1½.) Arenig Series; near Shelve, west Salop. RANGE: Arenig Series. [Syn., *Lingula plumbea, Obolella plumbea.*]

5. **Heterorthis retrorsistria** (M'Coy). (×1.) Caradoc Series; near Moelfre, Dyfed. RANGE: Caradoc Series. [Syn., *Orthis retror-sistria.*]

6–8.* **Heterorthis alternata** (J. de C. Sowerby). (×1.) 6, dorsal valve, internal mould. 7,8, ventral valve, artificial cast and internal mould. Caradoc Series; Soudley, near Church Stretton, Salop. RANGE: Caradoc Series. [Syn., *Orthis alternata.*]

9, 10. **Dinorthis flabellulum** (J. de C. Sowerby). Internal moulds of ventral and dorsal valves (×1¼.) Caradoc Series; Coston, near Clunbury, Salop. RANGE: Caradoc Series. [Syn., *Orthis flabellulum.*]

11–13. **Dalmanella horderleyensis** (Whittington). (×1½.) 11, artificial cast of ventral valve. 12,13, internal moulds of ventral and dorsal valves. Caradoc Series; Horderley, Salop. RANGE: Genus, Ordovician, Llanvirn Series–Silurian; Species, Caradoc Series. [Syn., *Wattsella horderleyensis.*]

Plate 4

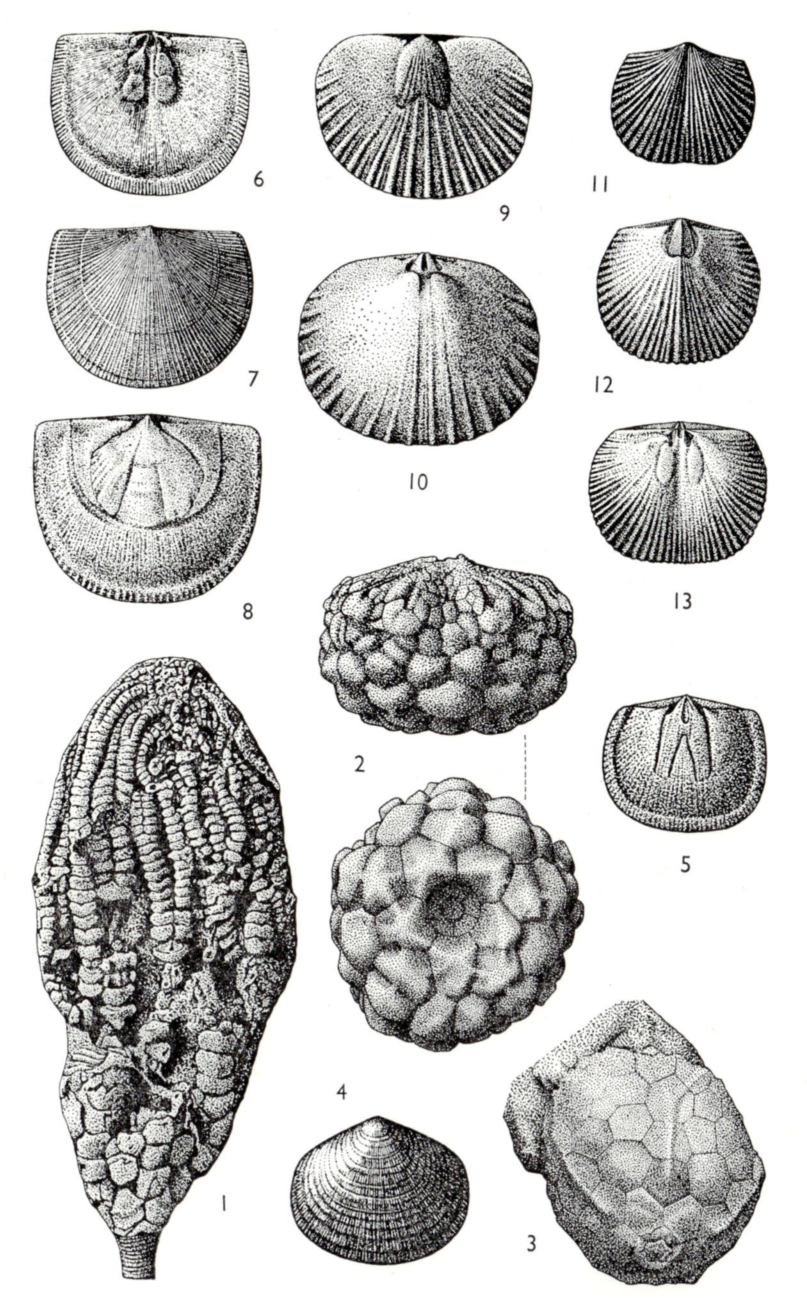

Plate 5

Ordovician Brachiopods

1. **Christiania perrugata** (Reed). Interior of dorsal valve (×1½.) Caradoc Series; Ardmillan, Girvan, Ayrshire. RANGE: Genus, Caradoc–Ashgill Series; Species, Caradoc Series.

2.* **Onniella broeggeri** Bancroft. (×2½.) Caradoc Series; River Onny Valley, near Wistanstow, Salop. RANGE: Genus, Caradoc–Ashgill Series; Species, Caradoc Series.

3, 4. **Nicolella actoniae** (J. de C. Sowerby). (×1½.) 3, Internal mould of ventral valve; 4, exterior of dorsal valve. Caradoc Series; near Acton Scott, Salop. RANGE: Genus, Caradoc–Ashgill Series; Species, Caradoc Series. [Syn., *Orthis actoniae.*]

5, 6.* **Sowerbyella sericea** (J. de C. Sowerby). Internal mould and exterior of ventral valve (×2.) Caradoc Series; 5, near Horderley, Salop. 6, Soudley, Salop. RANGE: Genus, Llanvirn–Ashgill Series; Species, Caradoc Series.

7, 8.* **Macrocoelia expansa** (J. de C. Sowerby). 7, ventral valve (×1¼.) 8, internal mould of ventral valve (×¾.) Caradoc Series; near Welshpool, Powys. RANGE: Genus, Llandeilo–Caradoc Series; Species, Caradoc Series. [Syn., *Rafinesquina expansa, Strophomena expansa.*]

9, 10.* **Strophomena grandis** (J. de C. Sowerby). 9, Internal mould of ventral valve, and 10, of dorsal valve (×1.) Caradoc Series; near Cheney Longville, Salop. RANGE: Caradoc Series. [Syn., *Longvillia grandis, Orthis grandis.*]

Plate 5

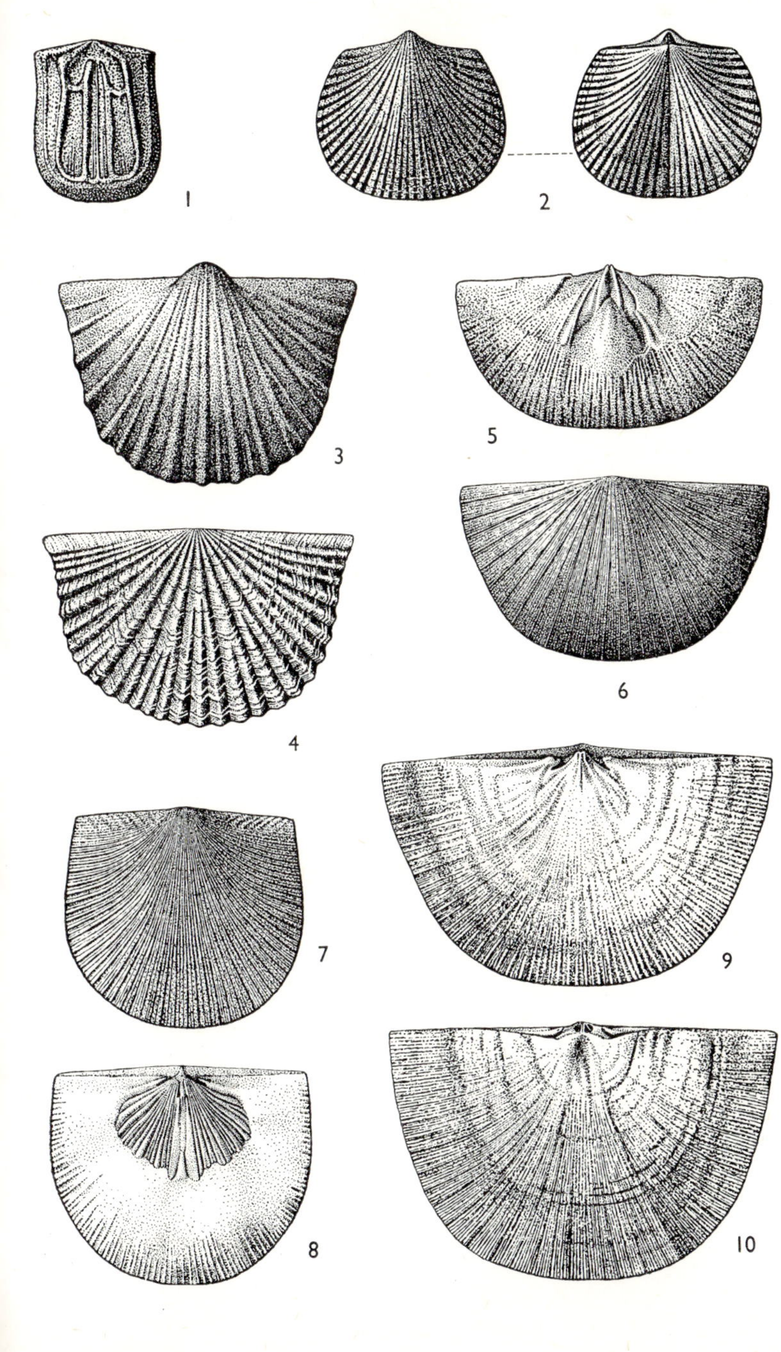

Plate 6

Cambrian Hyolithid (Fig. 11) and Ordovician Brachiopods (Figs. 1–9), Gastropod (Fig. 10), and Bivalves (Figs. 12, 13)

1, 2. **Sampo ruralis** (Reed). (×1½.). Ashgill Series; Thraive Glen, Girvan, Ayrshire. RANGE: Genus, Caradoc–Ashgill Series; Species, Ashgill Series. [Syn., *Plectambonites ruralis.*]

3, 4. **Reuschella horderleyensis** Bancroft. Internal moulds of ventral and dorsal valves. (×1.) Caradoc Series; Horderley, Salop. RANGE: Caradoc Series.

5. 6. **Hirnantia sagittifera** (M'Coy). Both valves and internal mould of dorsal valve. (×1.) Ashgill Series; near Knock, Cumbria. RANGE: Ashgill Series. [Syn., *Orthis sagittifera.*]

7–9.* **Harknessella vespertilio** (J. de C. Sowerby). (×1¼.) 7,8. Internal moulds of dorsal and ventral valves. 9, dorsal valve. Caradoc Series; Coston, near Clunbury, Salop. RANGE: Genus, Llandeilo–Caradoc Series; Species, Caradoc Series. [Syn., *Orthis vespertilio.*]

10.* **Cyclonema longstaffae** Lamont. (×1.) Ashgill Series: Shalloch Mill, Girvan, Ayrshire. RANGE: Genus, Caradoc Series–Upper Silurian; Species, Ashgill Series.

11. **Hyolithes magnificus** Stubblefield & Bulman. (×1.) Tremadoc Series; Cherme's Dingle, near the Wrekin, Salop. RANGE: Genus, Lower Cambrian–Upper Silurian; Species, Tremadoc Series.

12.* **Modiolopsis orbicularis** (J. de C. Sowerby). (×¾.) Caradoc Series; Hatton, Salop. RANGE: Genus, Ordovician, Llandeilo Series–Silurian (Downton Series); Species, Caradoc Series, [Syn., *Ambonychia orbicularis.*]

13.* **Byssonychia radiata** (Hall). (×1.) Ashgill Series; Girvan, Ayrshire. RANGE: Genus, Ordovician, Caradoc Series–Silurian; Species, Caradoc–Ashgill Series.

Plate 6

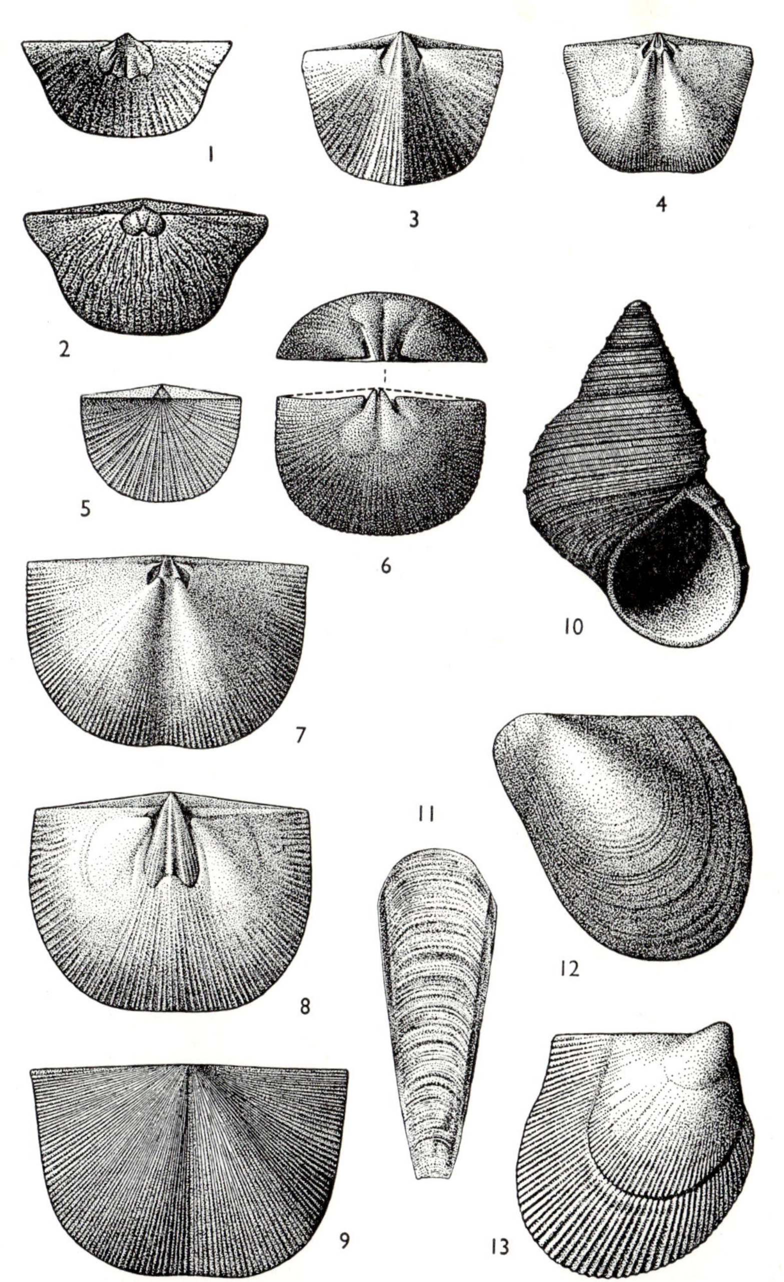

Plate 7

Cambrian Trilobites (Figs. 4–8) and Ordovician Monoplacophorans (Figs. 1, 2), Cephalopod (Fig. 3)

1.* **Sinuites subrectangularis** Reed. (×1.) Ashgill Series; Girvan, Ayrshire. RANGE: Genus, Ordovician, Caradoc Series–Upper Silurian; Species, Ashgill Series.

2.* **Cyrtolites nodosus** (Salter). (×2.) Caradoc Series; Cheney Longville, Salop. RANGE: Genus, Ordovician, Llandeilo Series–Lower Silurian; Species, Caradoc Series. [Syn., *Bellerophon nodosus*.]

3.* **'Orthoceras' vagans** Salter. (×1½.) Ashgill Series; near Bala, Gwynedd. RANGE: Ashgill Series.

4. **Geragnostus callavei** (Lake). (×3.) Tremadoc Series; Sheinton, Salop. RANGE: Genus, Tremadoc–Caradoc Series; Species, Tremadoc Series. [Syn., *Agnostus callavei*.]

5. **Shumardia pusilla** (Sars). (×6.) Tremadoc Series; Sheinton, Salop. RANGE: Genus, Tremadoc–Ashgill Series; Species, Tremadoc Series.

6.* **Asaphellus homfrayi** (Salter). (×¾.) Tremadoc Series; Sheinton, Salop. RANGE: Tremadoc Series. [Syn., *Asaphus homfrayi*.]

7. **Euloma monile** Salter. Cephalon (×6.) Tremadoc Series; Sheinton, Salop. RANGE: Tremadoc Series.

8.* **Parabolinella triarthra** (Callaway). (×1.) Tremadoc Series; Sheinton, Salop. RANGE: Tremadoc Series. [Syn., *Olenus triarthrus*.]

Plate 7

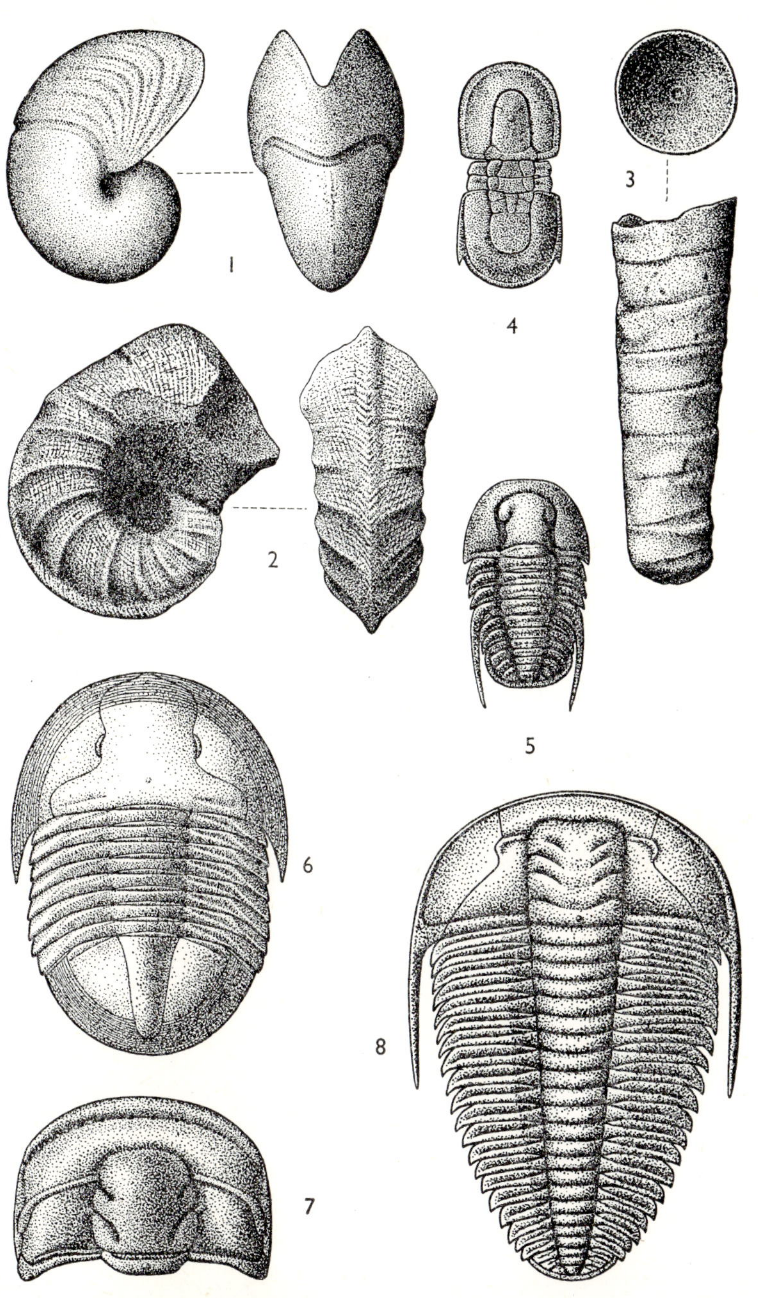

Plate 8

Cambrian and Ordovician Trilobites

1. **Stapeleyella inconstans** Whittard. (×3.) Llanvirn Series; Linley, Salop. RANGE: Llanvirn Series. [Syn., *Trinucleus murchisoni* of authors, in part.]

2, 3. **Placoparina sedgwicki** (M'Coy) **shelvensis** Hughes. Cephalon and pygidium (×1½.) Llanvirn Series; Linley, Salop. RANGE: Llanvirn Series. [Syn., *Cheirurus sedgwicki, Eccoptochile sedgwicki.*]

4. **Ampyx linleyensis** Whittard. (×1½.) Llanvirn Series; Linley, Salop. RANGE: Genus, Llanvirn–Caradoc Series; Species, Llanvirn Series. [Syn., *Ampyx mammillatus* subsp. *austini* of authors.]

5. **Placoparia cambriensis** Hicks. (×4.) Llanvirn Series; Ritton Castle, Salop. RANGE: Genus, Arenig–Llandeilo Series; Species, Arenig–Llanvirn Series. [Syn., *Placoparia zippei.*]

6.* **Angelina sedgwicki** Salter. (×1.) Tremadoc Series; Tremadoc, Gwynedd. RANGE: Tremadoc Series.

7. **Megalaspidella ? murchisonae** (Murchison). (×1.) Arenig Series; Shelve, Salop. RANGE: Genus, Tremadoc–Arenig Series; Species, Arenig Series. [Syn., *Niobella selwyni, Ogygia selwyni.*]

8. **Neseuretus ramseyensis** (Hicks). Cranidium (×1.) Arenig Series; near Shelve, Salop. RANGE: Genus, Arenig–Llanvirn Series; Species, Arenig Series. [Syn., *Calymene murchisoni, Synhomalonotus murchisoni.*]

Plate 8

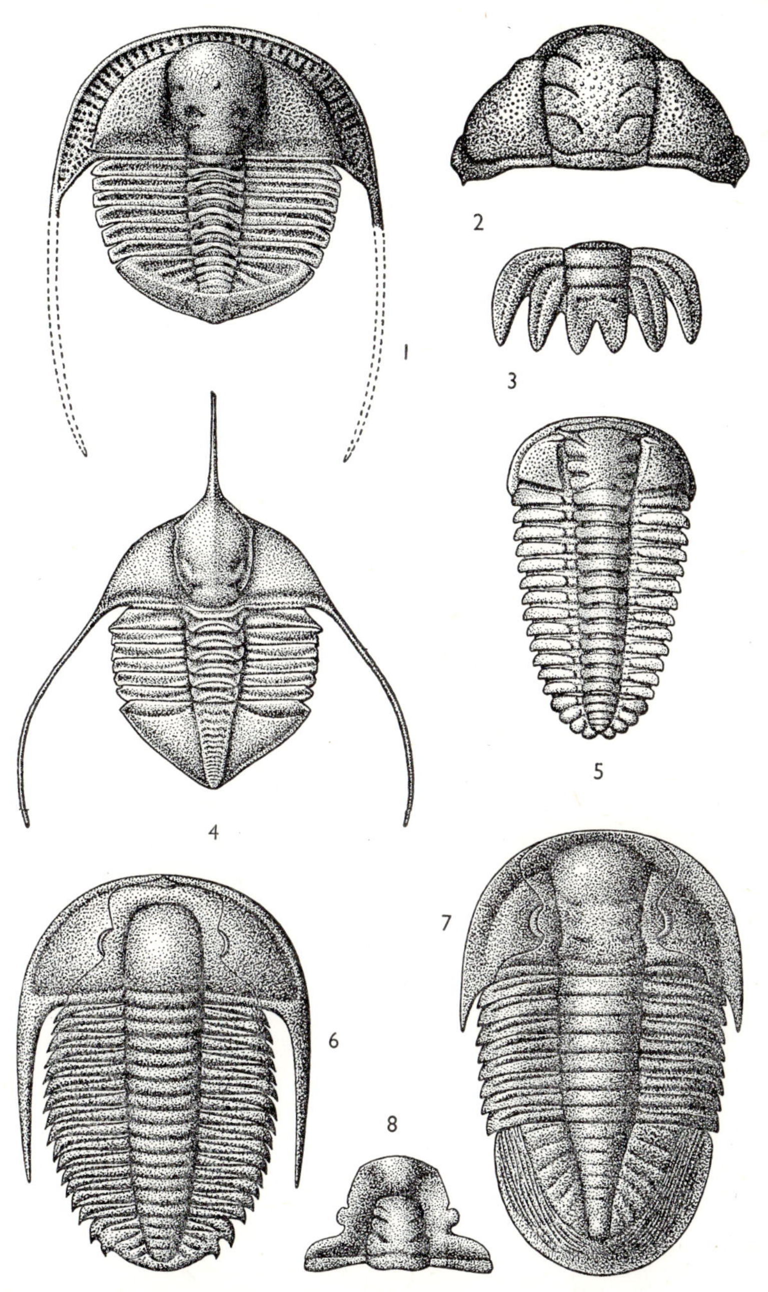

Plate 9

Ordovician Trilobites

1, 2.* **Flexicalymene cambrensis** (Salter). Llandeilo Series; 1, cephalon (×1½), Llandilo, Dyfed. 2, pygidium (×1), Meadowtown, Salop. RANGE: Genus, Ordovician, Llandeilo Series–Lower Silurian; Species, Llandeilo Series. [Syn., *Calymene cambrensis.*]

3, 4.* **Platycalymene duplicata** (Murchison). Caradoc Series. 3, cephalon (×1½), Gwernyffyd, Powys; 4, pygidium (×1¼), Llandrindod Wells, Powys. RANGE: Genus, Llanvirn–Caradoc Series; Species, Caradoc Series. [Syn., *Calymene duplicata.*]

5. **Marrolithus favus** (Salter). Cephalon (×3.) Llandeilo Series; Llandilo, Dyfed. RANGE: Genus, Llanvirn–Caradoc Series; Species, Llandeilo Series. [Syn., *Trinucleus favus.*]

6, 7. **Encrinuroides sexcostatus** (Salter). Cephalon and pygidium (×1.) Ashgill Series; near Haverfordwest, Dyfed. RANGE: Ashgill Series. [Syn., *Encrinurus sexcostatus.*]

8. **Selenopeltis inermis** (Beyrich). (×1½.) Llanvirn Series; Llanvirn, Dyfed. RANGE: Llanvirn–Llandeilo Series. [Syn., *Acidaspis buchi* (Barrande), *Selenopeltis buchi.*]

9. **Flexicalymene caractaci** (Salter). (×1.) Caradoc Series; Marshbrook, Salop. RANGE: Genus, Ordovician, Llandeilo Series–Lower Silurian; Species, Caradoc Series. [Syn., *Calymene caractaci.*]

Plate 9

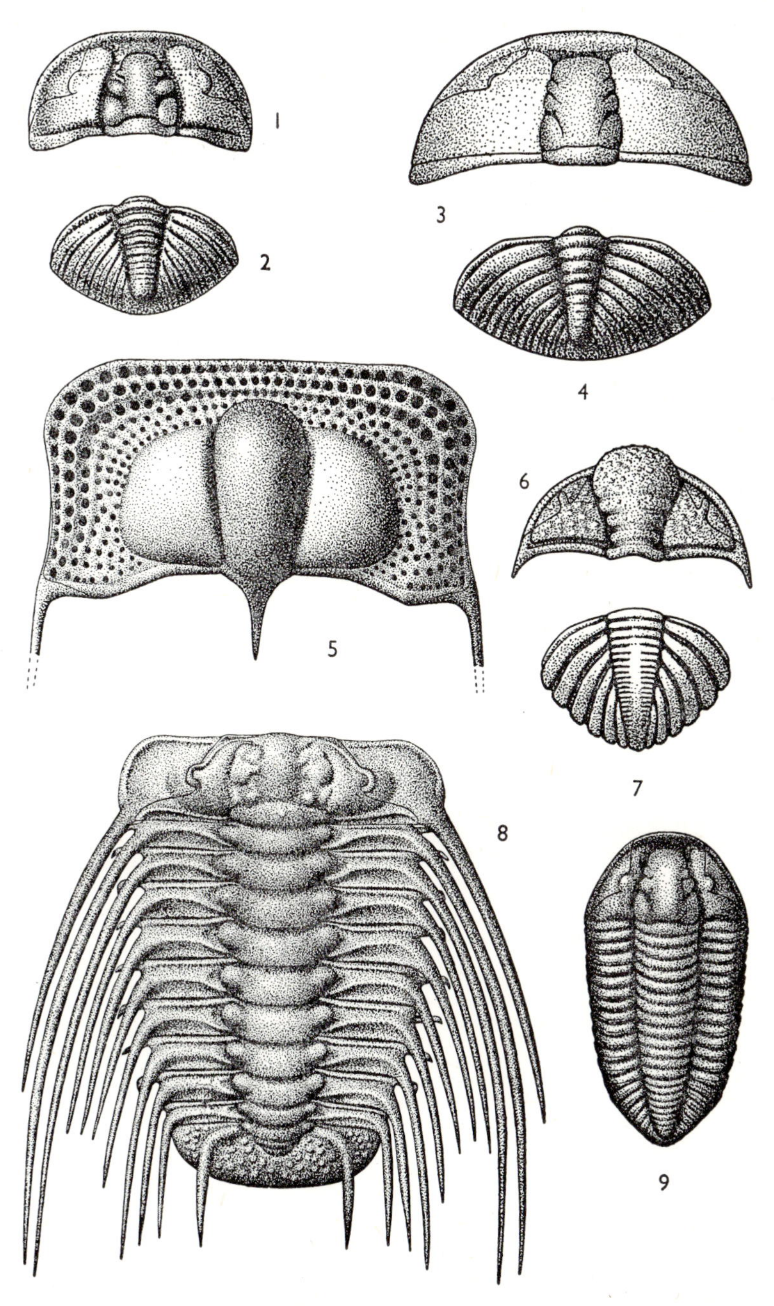

Plate 10

Ordovician Trilobites

1.* **Trinucleus fimbriatus** Murchison. (×2.) Caradoc Series; Builth Wells, Powys. RANGE: Genus, Llanvirn–Caradoc Series; Species, Caradoc Series.

2. **Remopleurides girvanensis** Reed. (×1½.) Caradoc Series; Girvan, Ayrshire. RANGE: Genus, Caradoc–Ashgill Series; Species, Caradoc Series.

3.* **Cnemidopyge bisecta** (Elles). (×1½.) Caradoc Series; Builth Wells, Powys. RANGE: Llandeilo Series. [Syn., *Ampyx nudus*, *Cnemidopyge nuda*.] Genus, Llandeilo–Caradoc, Species, Caradoc Series.

4. **Salterolithus caractaci** (Murchison). (×1¼.) Caradoc Series; Welshpool, Powys. RANGE: Caradoc Series. [Syn., *Trinucleus caractaci*, *Trinucleus concentricus* of authors in part, *Trinucleus intermedius* Wade.]

5.* **Onnia gracilis** (Bancroft). (×1¼.) Caradoc Series; River Onny Valley, near Wistanstow, Salop. RANGE: Caradoc Series. [Syn., *Trinucleus concentricus* of authors in part, *Cryptolithus gracilis*.]

6.* **Ogygiocarella angustissima** (Salter). (×¾.) Caradoc Series; Gwernyffyd, Powys. RANGE: Llandeilo–Caradoc Series. [Syn., *Ogygia buchi* of authors, *Ogygiocaris buchi*.]

7.* **Basilicus tyrannus** (Murchison). (×½.) Llandeilo Series; Llandilo, Dyfed. RANGE: Llandeilo Series. [Syn., *Asaphus tyrannus*.]

Plate 10

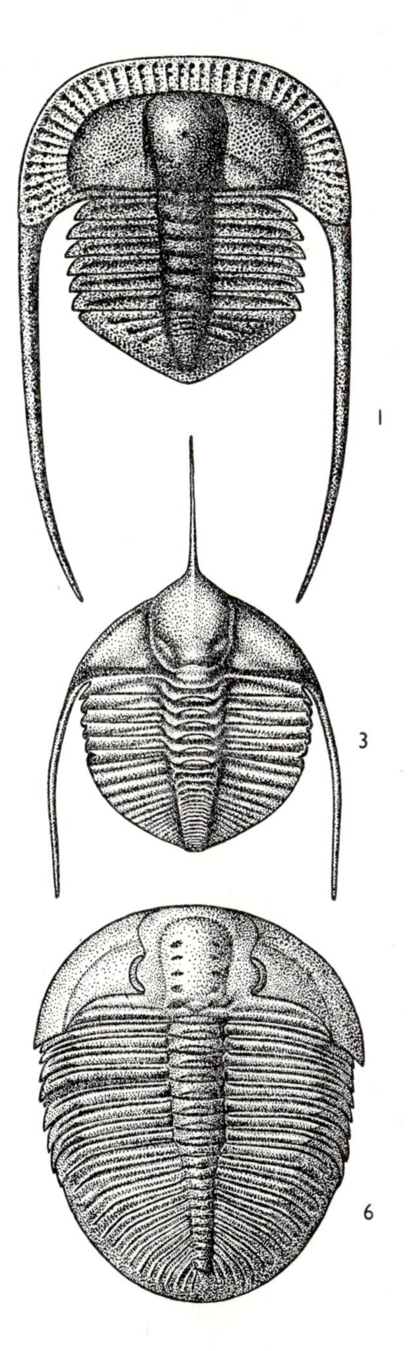

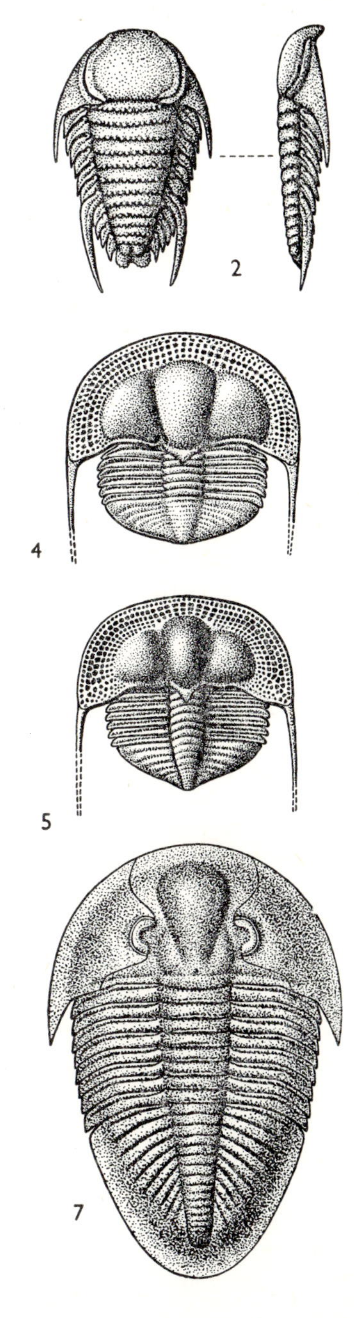

Plate 11

Ordovician Trilobites

1, 2. **Chasmops extensa** (Boeck). (×¾.) 1, cephalon. 2, pygidium. Caradoc Series; River Onny, near Wistanstow, Salop. RANGE: Genus, Caradoc–Ashgill Series; Species, Caradoc Series.

3–5. **Broeggerolithus broeggeri** (Bancroft). (×2.) 3,4, ventral and dorsal views of cephalon. 5, pygidium. Caradoc Series; near Horderley, Salop. RANGE: Caradoc Series.

6, 7. **Brongniartella bisulcata** (M'Coy). 6, cephalon (×½.) 7, pygidium (×¾.) Caradoc Series; Marshbrook, Salop. RANGE: Caradoc Series. [Syn., *Homalonotus bisulcatus.*]

8, 9. **Kloucekia apiculata** (M'Coy). (×1½.) 8, cephalon. 9, pygidium. Caradoc Series; Horderley, Salop. RANGE: Caradoc Series. [Syn., *Acaste apiculata, Phacopidina apiculata, Phacops apiculatus.*]

10, 11.* **Diacalymene drummuckensis** (Reed). 10, cephalon (×1½.) 11, pygidium (×2). Ashgill Series; Girvan, Ayrshire. RANGE: Genus, Ordovician, Caradoc Series–Silurian; Species, Ashgill Series. [Syn., *Calymene blumenbachi* var. *drummuckensis.*]

12. **Cybeloides girvanensis** (Reed). (×1.) Ashgill Series; Girvan, Ayrshire. RANGE: Genus, Caradoc–Ashgill Series; Species, Ashgill Series. [Syn., *Cybeloides loveni* var. *girvanensis.*]

Plate 11

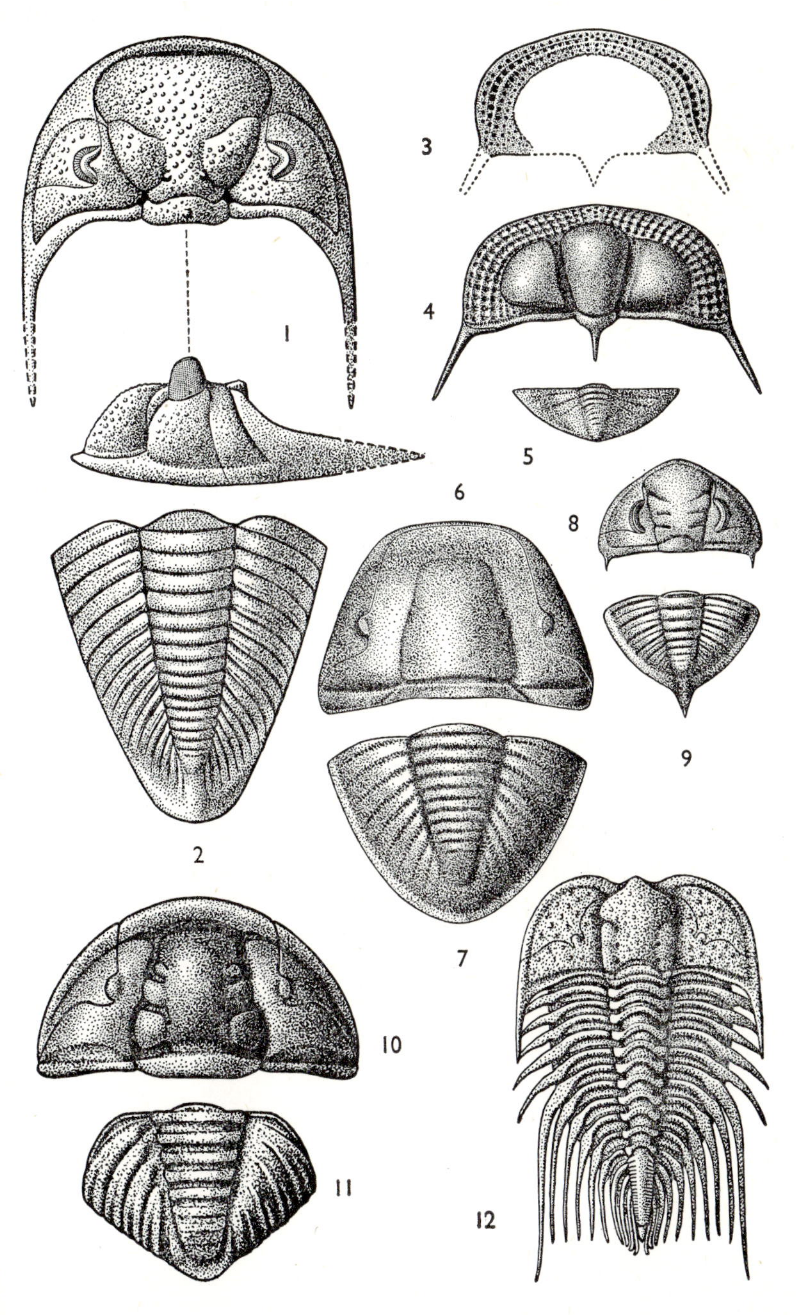

Plate 12

Ordovician Trilobites (Figs. 1–10), Ostracods (Figs. 11, 12) and Rostroconch (Fig. 13)

1.* **Paraproetus girvanensis** (Nicholson & Etheridge). ($\times 1\frac{1}{2}$.) Ashgill Series; Girvan, Ayrshire. RANGE: Genus, Ordovician–Devonian; Species, Ashgill Series.

2. **Flexicalymene quadrata** (King). ($\times 1\frac{1}{2}$.) Ashgill Series; Blaen-y-Cwm Valley, Berwyn Hills, Powys. RANGE: Genus, Llandeilo Series–Lower Silurian; Species, Ashgill Series. [Syn., *Calymene quadrata*.]

3. **Tretaspis sortita** (Reed). ($\times 2$.) Ashgill Series; near Girvan, Ayrshire. RANGE: Genus, Caradoc–Ashgill Series; Species, Ashgill Series. [Syn., *Trinucleus ceriodes* var. *sortita*.]

4.* **Pseudosphaerexochus octolobatus** (M'Coy). ($\times \frac{3}{4}$.) Ashgill Series; Girvan, Ayrshire. RANGE: Genus, Caradoc–Ashgill Series; Species, Ashgill Series. [Syn., *Cheirurus octolobatus*.]

5. **Phillipsinella parabola** (Barrande). ($\times 2$.) Ashgill Series; Girvan, Ayrshire. RANGE: Ashgill Series.

6. **Sphaerocoryphe thomsoni** (Reed). ($\times 2$.) Ashgill Series; Girvan, Ayrshire. RANGE: Ashgill Series. [Syn., *Cheirurus thomsoni*.]

7, 8. **Kloucekia robertsi** (Reed). 7, cephalon ($\times 1\frac{1}{2}$.) 8, pygidium ($\times 1\frac{1}{4}$.) Ashgill Series; Haverfordwest, Dyfed. RANGE: Genus, Caradoc Series–Lower Silurian; Species, Ashgill Series. [Syn., *Phacops robertsi*.]

9, 10. **Corrugatagnostus sol** Whittard. ($\times 3$.) 9, cephalon. 10, pygidium. Ashgill Series; near Girvan, Ayrshire. RANGE: Genus, Llanvirn–Ashgill Series; Species, Ashgill Series. [Syn., *Agnostus perrugatus* of authors, in part.]

11. **Tallinnella scripta** (Harper). ($\times 10$.) Caradoc Series; near Cressage, Salop. RANGE: Genus, Llandeilo–Caradoc Series; Species, Caradoc Series. [Syn., *Tetradella scripta*.]

12. **Primitia maccoyi** Salter. ($\times 10$.) Ashgill Series; Chair of Kildare, Kildare, Ireland. RANGE: Genus, Llandeilo–Ashgill Series; Species, Ashgill Series.

13.* **Ribeiroidea lapworthi** (Etheridge). ($\times 2$.) Caradoc Series; Girvan, Ayrshire. RANGE: Caradoc Series.

Plate 12

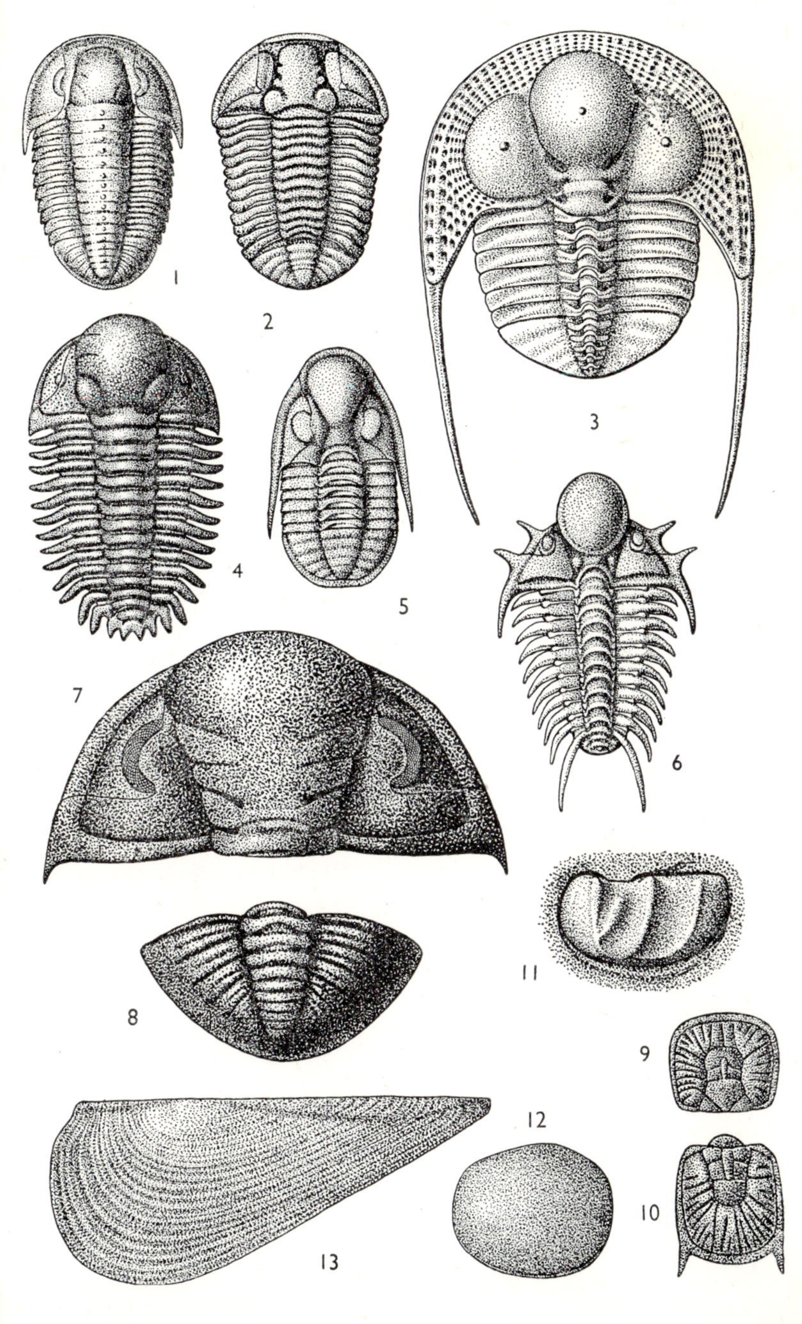

Plate 13

Cambrian and Ordovician Graptolites

1.* **Callograptus** cf. **salteri** Hall. (×1.) Ashgill Series; near Girvan, Ayrshire. RANGE: Genus, Upper Cambrian–Lower Carboniferous; Species, Ashgill Series.

2.* **Didymograptus hirundo** Salter. (×2.) Arenig Series; Skiddaw, Keswick, Cumbria. RANGE: Genus, Arenig–Llandeilo Series; Species, Arenig Series.

3. **Didymograptus extensus** (Hall). (×2.) Arenig Series; Lleyn, Gwynedd. RANGE: Genus, Arenig–Llandeilo Series; Species, Arenig Series.

4. **Glyptograptus teretiusculus** (Hisinger). (×2.) Llandeilo Series; near Pwllheli, Gwynedd. RANGE: Genus, Ordovician, Arenig Series–Silurian, Llandovery Series; Species, Llandeilo Series.

5. **Tetragraptus serra** (Brongniart). (×2.) Arenig Series; near Keswick, Cumbria. RANGE: Genus, Arenig–Llanvirn Series; Species, Arenig Series.

6. **Phyllograptus angustifolius** Hall. (×2.) Arenig Series; near Keswick, Cumbria. RANGE: Arenig Series.

7. **Ptilograptus acutus** (Hopkinson). (×1½.) Arenig Series; Shelve, Salop. RANGE: Genus, Lower Ordovician–Upper Silurian; Species, Arenig Series.

8.* **Dictyonema flabelliforme** (Eichwald). (×1.) Tremadoc Series; near Ffestiniog, Gwynedd. RANGE: Genus, Upper Cambrian–Lower Carboniferous; Species, Tremadoc–Arenig Series.

9.* **Clonograptus tenellus** (Linnarsson). (×2.) Tremadoc Series; Cherme's Dingle, near The Wrekin, Salop. RANGE: Tremadoc–Arenig Series.

10. **Dichograptus octobrachiatus** (Hall). (×2.) Arenig Series; near Keswick, Cumbria. RANGE: Genus, Arenig–Llanvirn Series; Species, Arenig Series.

Plate 13

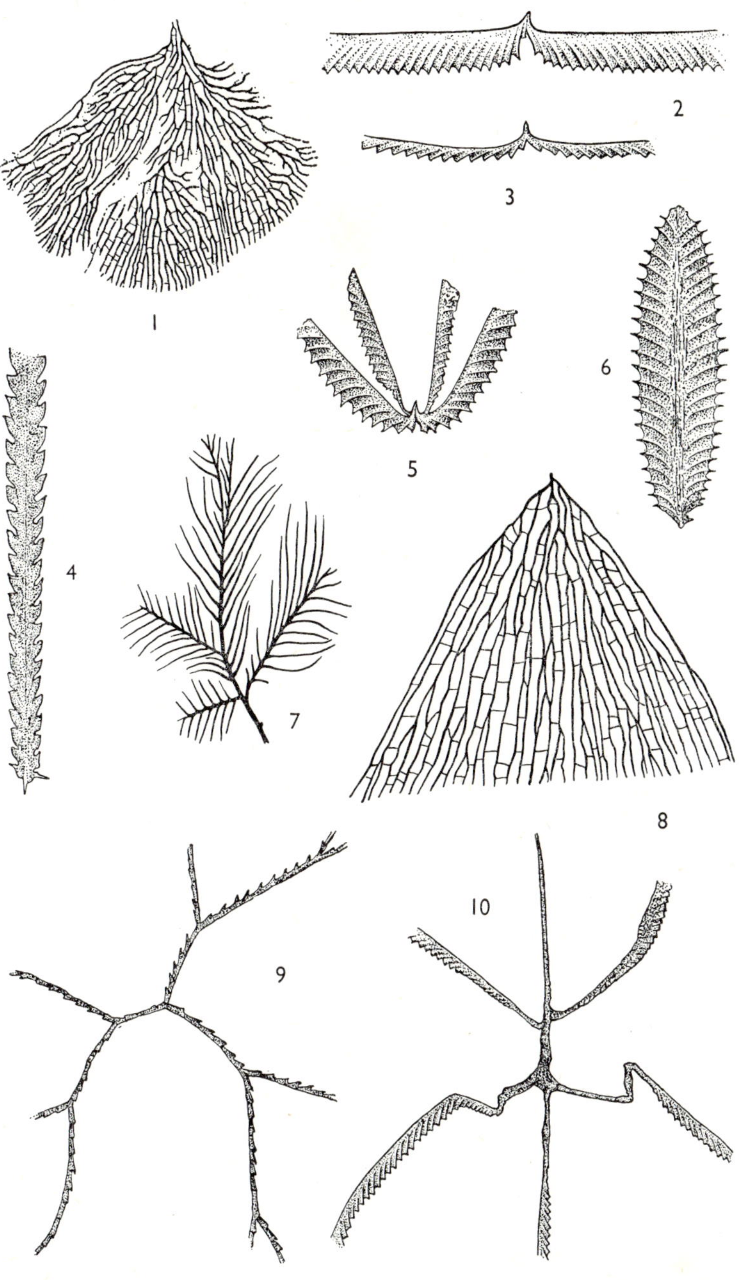

Plate 14

Ordovician Graptolites (Figs. 1–12) and Conodont (Fig. 13)

1. **Leptograptus flaccidus** (Hall). (×2.) Caradoc Series; near Moffat' Dumfriesshire. RANGE: Genus, Arenig–Caradoc Series; Species, Caradoc Series.
2. **Pleurograptus linearis** (Carruthers). (×2.) Caradoc Series; near Moffat, Dumfriesshire. RANGE: Caradoc Series.

3.* **Dicranograptus clingani** Carruthers. (×2.) Caradoc Series; near Moffat, Dumfriesshire. RANGE: Genus, Llandeilo–Caradoc Series; Species, Caradoc Series.

4.* **Didymograptus murchisoni** (Beck). (×1.) Llanvirn Series; Abereiddy Bay, Dyfed. RANGE: Genus, Arenig–Llandeilo Series; Species, Llanvirn Series.

5. **Didymograptus bifidus** Hall. (×2.) Llanvirn Series; near Aberdaron, Gwynedd. RANGE: Genus, Arenig–Llandeilo Series; Species, Llanvirn Series.

6. **Nemagraptus gracilis** (Hall). (×2.) Caradoc Series; near Moffat, Dumfriesshire. RANGE: Llandeilo–Caradoc Series.

7.* **Diplograptus multidens** (Elles). (×2.) Caradoc Series; near Haverfordwest, Dyfed. RANGE: Genus, Ordovician, Llanvirn Series–Silurian, Llandovery Series; Species, Caradoc Series. [Syn., *Mesograptus multidens*.]

8.* **Climacograptus bicornis** (Hall). (×2.) Caradoc Series; Moffat, Dumfriesshire. RANGE: Genus, Ordovician, Arenig Series–Silurian, Llandovery Series; Species, Caradoc Series.

9. **Orthograptus truncatus** (Lapworth). (×2.) Caradoc Series; near Girvan, Ayrshire. RANGE: Genus, Ordovician, Caradoc Series–Silurian, Llandovery Series; Species, Caradoc Series.

10.* **Orthograptus calcaratus** (Lapworth). (×2.) Caradoc Series; Conway, Gwynedd. RANGE: Genus, Ordovician, Caradoc Series–Silurian, Llandovery Series; Species, Caradoc Series.

11. **Climacograptus wilsoni** Lapworth. (×2.) Caradoc Series; near Moffat, Dumfriesshire. RANGE: Genus, Ordovician, Arenig Series–Silurian, Llandovery Series; Species, Caradoc Series.

12.* **Dicellograptus anceps** Nicholson. (×2.) Ashgill Series; near Moffat, Dumfriesshire. RANGE: Genus, Llanvirn–Ashgill Series; Species, Ashgill Series.

13. **Trichonodella flexa** Rhodes. (×40.) Llandeilo Series; near Llanfihangel Aberbythych, Dyfed. RANGE: Genus, Middle Ordovician–Upper Silurian; Species, Llandeilo Series.

Plate 14

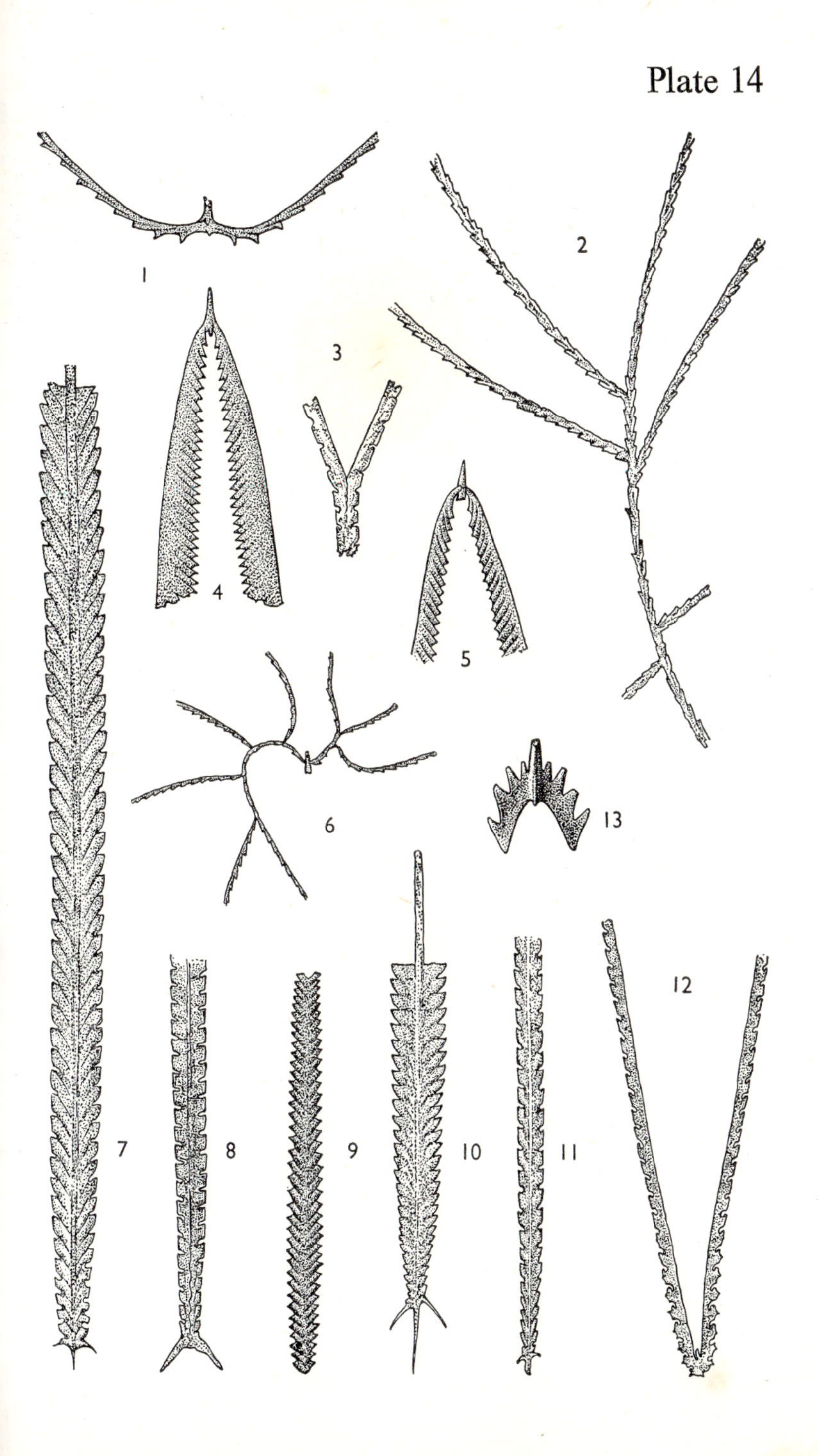

Plate 15

Silurian Corals (Figs. 1–6), Sponges (Figs. 7, 8) and Alga (Fig. 9)

1.* **Favosites gothlandicus** Lamarck forma **forbesi** (Edwards & Haime). 1 (×¾), 1*a* (×1½.) Wenlock Series; Dudley, West Midlands. RANGE: Genus, Upper Ordovician–Upper Devonian (? Trias); Species, Wenlock–Ludlow Series.

2.* **Acervularia ananas** (Linné). (×2.) Wenlock Series; Dudley, West Midlands. RANGE: Genus, Silurian; Species, Wenlock Series.

3.* **Halysites catenularius** (Linné). (×4.) Wenlock Series; Dorrington, near Hereford. RANGE: Genus, Ordovician–Silurian; Species, Wenlock Series.

4.* **Arachnophyllum murchisoni** (Edwards & Haime). 4 (×¾), 4*a* (×1½.) Wenlock Series; Dudley, West Midlands. RANGE: Genus, Llandovery–Wenlock Series; Species, Wenlock Series. [Syn., *Strombodes murchisoni*, *Strombodes phillipsi* (Orbigny).]

5. **Thamnopora cristata** (Blumenbach). 5 (×1), 5*a* (×3.) Wenlock Series; Dudley, West Midlands. RANGE: Genus, Silurian–Permian; Species, Wenlock Series. [Syn., *Favosites cristata, Pachypora cristata.*]

6.* **Tryplasma loveni** (Edwards & Haime). 6 (×1), 6*a* (×1½.) Wenlock Series; Dudley, West Midlands. RANGE: Genus, Silurian–Lower Devonian; Species, Wenlock Series. [Syn., *Cyathophyllum*? *loveni*, *Pholidophyllum loveni.*]

7.* **Ischadites koenigi** Murchison. (×1½.) Wenlock Series; Dudley, West Midlands. RANGE: Genus, Ordovician–Silurian; Species, Ordovician, Llandeilo Series–Silurian, Ludlow Series.

8.* **Amphispongia oblonga** Salter. (×¾.) Ludlow Series; Pentland Hills, Midlothian. RANGE: Genus, Silurian; Species, Ludlow Series.

9. **Mastopora fava** (Salter). (×1.) Llandovery Series; Mulloch Hill, Girvan, Ayrshire. RANGE: Genus, Ordovician–Silurian; Species, Llandovery Series. [Syn., *Nidulites favus.*]

Plate 15

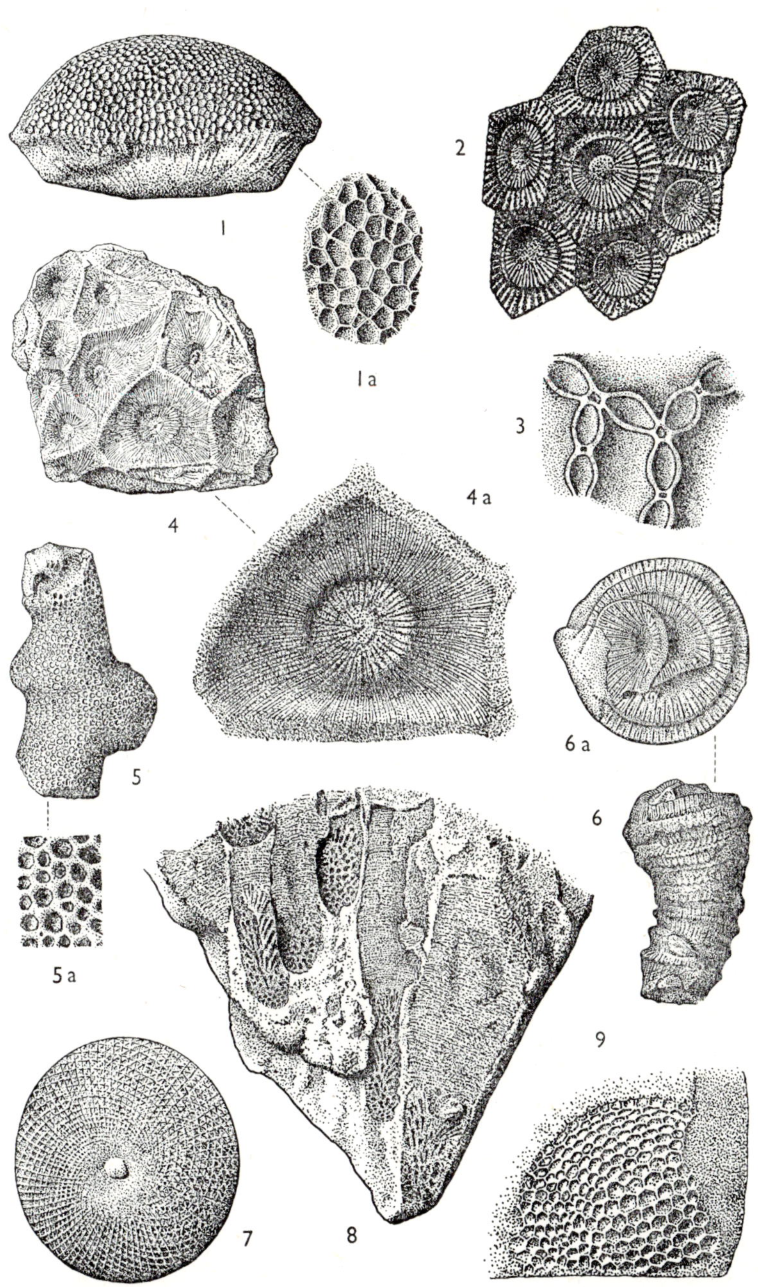

Plate 16

Silurian Bryozoan (Fig. 1), Corals (Figs. 2–9) and Hydrozoan (Fig. 10)

1. **Favositella interpuncta** (Quenstedt). 1 (×½), 1*a* (×4.) Wenlock Series; Dudley, West Midlands. RANGE: Genus, Ordovician–Silurian; Species, Wenlock Series.

2.* **Goniophyllum pyramidale** (Hisinger). (×1.) Wenlock Series; Dudley, West Midlands. RANGE: Genus, Llandovery–Wenlock Series; Species, Wenlock Series.

3.* **Ketophyllum subturbinatum** (Orbigny). (×½.) Wenlock Series; Dudley, West Midlands. RANGE: Genus, Llandovery–Wenlock Series; Species, Wenlock Series. [Syn., *Omphyma subturbinata*.]

4. **Rhabdocyclus fletcheri** (Edwards & Haime). (×2.) Wenlock Series; Dudley, West Midlands. RANGE: Genus, Silurian; Species, Wenlock Series. [Syn., *Palaeocyclus fletcheri*.]

5.* **Heliolites interstinctus** (Linné). (×6.) Wenlock Series; Much Wenlock, Salop. RANGE: Genus, Silurian–Middle Devonian; Species, Wenlock Series.

6. **Syringopora bifurcata** Lonsdale. (×¾.) Wenlock Series; Much Wenlock, Salop. RANGE: Genus, Silurian–Carboniferous; Species, Wenlock Series.

7, 8.* **Kodonophyllum truncatum** (Linné). 7 (×2), 8 (×1.) Wenlock Series; near Much Wenlock, Salop. RANGE: Genus, Middle–Upper Silurian; Species, Wenlock Series. [Syn., *Cyathophyllum truncatum*.]

9. **Thecia swinderniana** (Goldfuss). (×5.) Wenlock Series; Dudley, West Midlands. RANGE: Genus, Silurian–Devonian; Species, Wenlock Series.

10. **Labechia conferta** (Lonsdale). (×1½.) Wenlock Series; Coalbrookdale, Salop. RANGE: Genus, Ordovician–Silurian; Species, Wenlock Series.

Plate 16

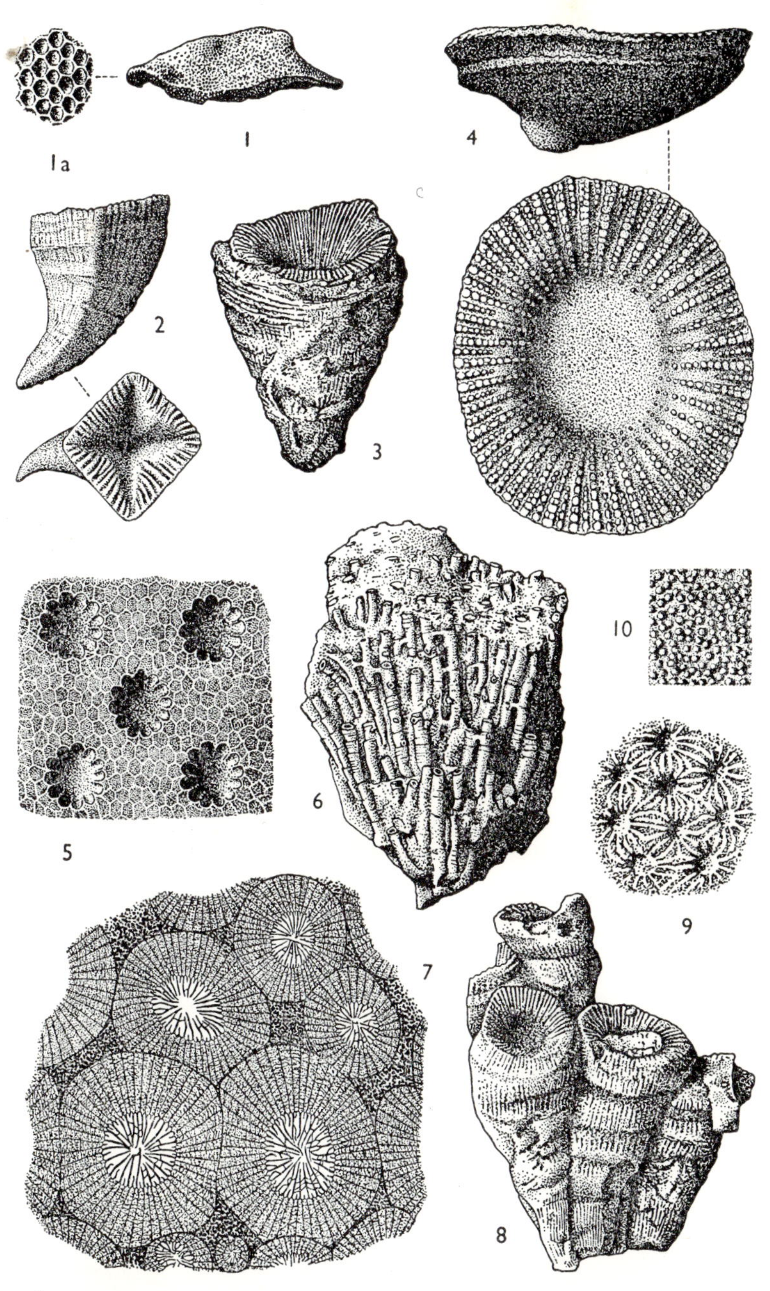

Plate 17

Silurian Worms (Figs. 1, 2, 6), Cricoconarida (Figs. 3–5) and Brachiopods (Figs. 7–12)

1.* **Serpulites longissimus** Murchison. ($\times \frac{3}{4}$.) Ludlow Series; Ludlow, Salop. RANGE: Genus, Ordovician–Carboniferous; Species, Ordovician, Ashgill Series–Silurian.

2. **Spirorbis tenuis** J. de C. Sowerby. ($\times$ 6.) Wenlock Series; Dudley, West Midlands. RANGE: Genus, Silurian–Recent; Species, Wenlock Series.

3. **Tentaculites scalaris** Schlotheim. ($\times 2\frac{1}{2}$.) Llandovery Series; Minsterley, Salop. RANGE: Genus, Ordovician–Devonian; Species, Silurian.

4.* **Tentaculites ornatus** J. de C. Sowerby. ($\times 2\frac{1}{2}$.) Wenlock Series; Dudley, West Midlands. RANGE: Genus, Ordovician–Devonian; Species, Wenlock Series.

5.* **Cornulites serpularius** Schlotheim. ($\times$ 1.) Wenlock Series; Dudley, West Midlands. RANGE: Genus, Ordovician–Silurian; Species, Wenlock–Ludlow Series.

6. **Keilorites squamosus** (Phillips). ($\times$ 1.) Ludlow Series; Salop. RANGE: Genus, Ordovician–Silurian; Species, Wenlock–Ludlow Series. [Syn., *Trachyderma squamosa.*]

7. **Skenidioides lewisii** (Davidson). ($\times$ 3.) Wenlock Series; Buildwas, Salop. RANGE: Genus, Ordovician, Ashgill Series–Silurian; Species, Llandovery–Wenlock Series. [Syn., *Skenidium lewisi.*]

8, 9. **Resserella canalis** (J. de C. Sowerby). ($\times$ 2.) Wenlock Series; Dudley, West Midlands. RANGE: Genus, Ordovician–Silurian; Species, Silurian. [Syn., *Parmorthis elegantula.*]

10, 11.* **Pentamerus oblongus** J. de C. Sowerby. ($\times \frac{3}{4}$.) Llandovery Series; Norbury, Salop. RANGE: Llandovery Series.

12. **Costistricklandia lirata** (J. de C. Sowerby). ($\times \frac{3}{4}$.) Llandovery Series; Ledbury, Herefordshire. RANGE: Llandovery Series. [Syn., *Pentamerus liratus, Stricklandinia lirata.*]

Plate 17

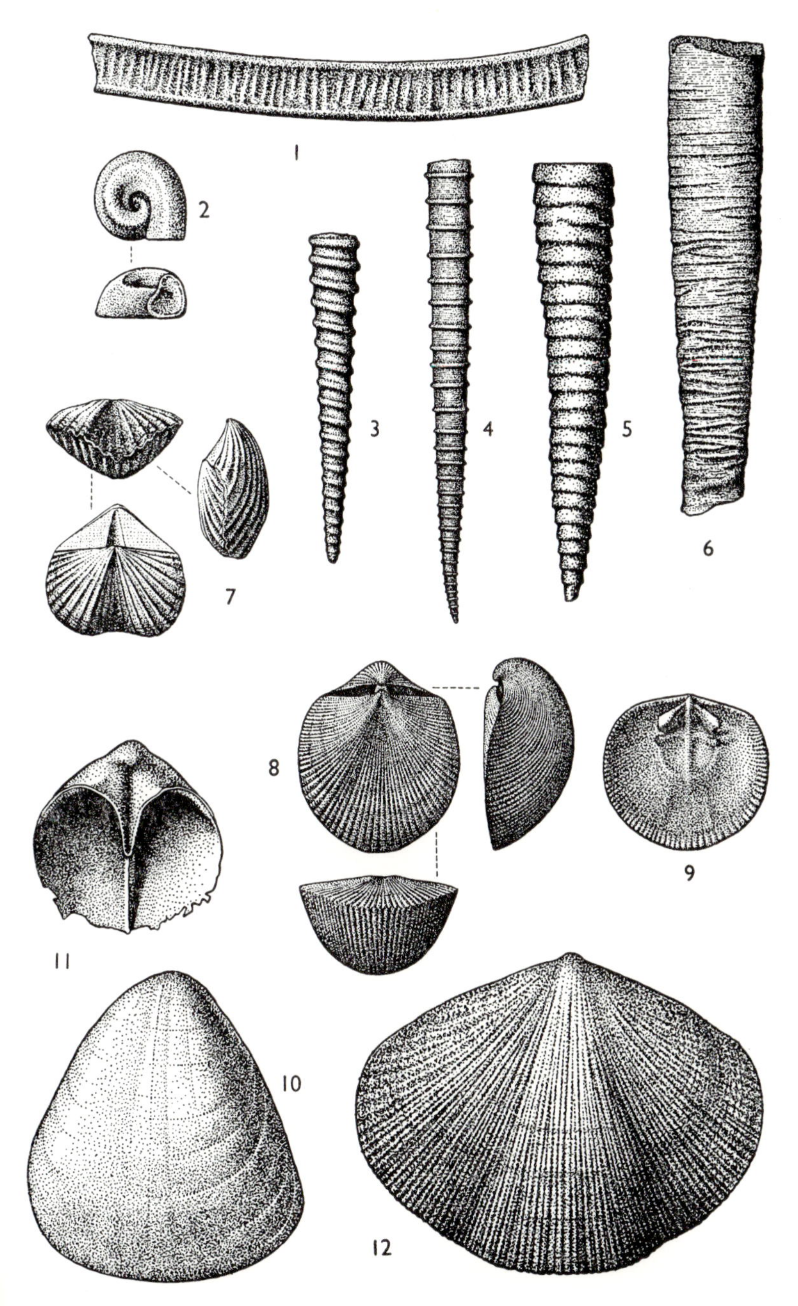

Plate 18

Silurian Brachiopods

1. **Rhynchotreta cuneata** (Dalman). (×1.) Wenlock Series; Dudley, West Midlands. RANGE: Genus, Silurian; Species, Wenlock Series. [Syn., *Rhynchonella cuneata.*]

2. **Gypidula galeata** (Dalman). (×1.) Wenlock Series; near Much Wenlock, Salop. RANGE: Genus, Silurian–Devonian; Species, Wenlock Series. [Syn., *Pentamerus dudleyensis.*]

3. **Anastrophia deflexa** (J. de C. Sowerby). (×1½.) Wenlock Series; Dudley, West Midlands. RANGE: Genus, Silurian–Devonian; Species, Wenlock Series. [Syn., *Rhynchonella deflexa.*]

4. **Dicoelosia biloba** (Linnaeus). (×3½.) Wenlock Series; Dudley, West Midlands. RANGE: Genus, Ordovician, Ashgill Series–Lower Devonian; Species, Wenlock–Ludlow Series. [Syn., *Bilobites biloba, Orthis biloba.*]

5–7. **Dolcrorthis rustica** (J. de C. Sowerby). Wenlock Series. 5 (×1.) Dudley, West Midlands; 6 (×2), interior of dorsal valve, 7 (×1½), interior of ventral valve, near Coalbrookdale, Salop. RANGE: Genus, Ordovician–Silurian, Wenlock Series; Species, Wenlock Series. [Syn., *Hebertella rustica, Orthis rustica.*]

Plate 18

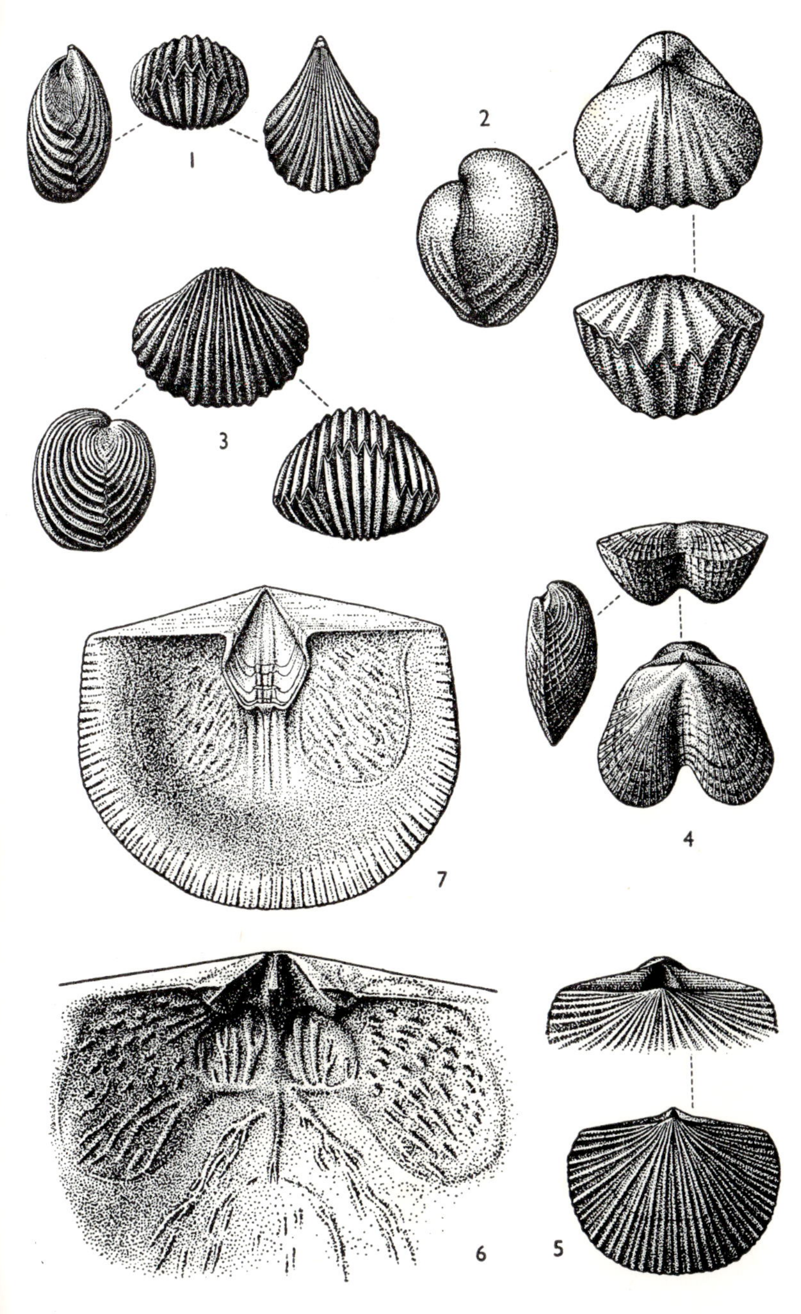

Plate 19

Silurian Brachiopods

1. **Trigonirhynchia stricklandii** (J. de C. Sowerby). (×1.) Wenlock Series; Malvern, Worcestershire. RANGE: Genus, Silurian; Species, Wenlock Series. [Syn., *Uncinulina stricklandi.*]

2. **Sphaerirhynchia wilsoni** (J. Sowerby). (×1½.) Wenlock Series; near Much Wenlock, Salop. RANGE: Genus, Silurian; Species, Wenlock–Ludlow Series. [Syn., *Rhynchonella wilsoni, Wilsonia wilsoni.*]

3, 4.* **Leptaena depressa** (J. de C. Sowerby). (×1.) Wenlock Series. 3, Usk, Gwent. 4, interior of ventral valve, Dudley, West Midlands. RANGE: Genus, Ordovician to Silurian; Species, Silurian. [Syn., *Leptaena rhomboidalis* of authors.]

5, 6.* **Strophonella euglypha** (Dalman). (×¾.) 5, dorsal valve, 6, interior of ventral valve. Wenlock Series; near Dudley, West Midlands. RANGE: Silurian. [Syn., *Strophomena euglypha.*]

7, 8. **Amphistrophia funiculata** (M'Coy). 7, dorsal valve (×2.) 8, interior of ventral valve (×1½). Wenlock Series; Dudley, West Midlands. RANGE: Genus, Silurian–Devonian; Species, Wenlock–Ludlow Series. [Syn., *Strophomena funiculata, Strophonella funiculata.*]

9. **Eoplectodonta duvalii** (Davidson). (×2.) Wenlock Series; Dudley, West Midlands. RANGE: Genus, Ordovician–Devonian; Species, Llandovery–Wenlock Series. [Syn., *Plectambonites transversalis.*]

Plate 19

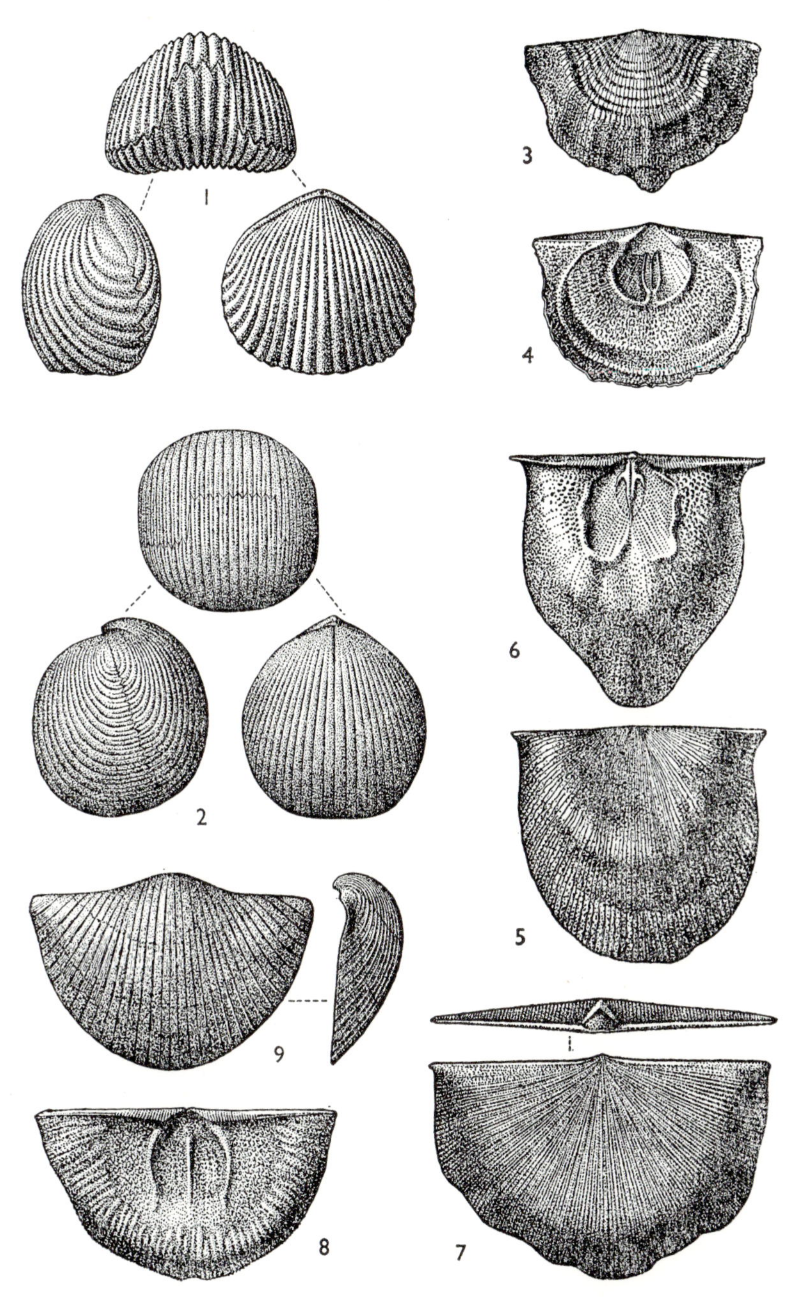

Plate 20

Silurian Brachiopods

1. **Atrypa reticularis** (Linnaeus). (×1.) Wenlock Series; Dudley, West Midlands. RANGE: Genus, Silurian–Devonian; Species, Silurian.

2. **Eospirifer radiatus** (J. de C. Sowerby). (×1.) Wenlock Series; Dudley, West Midlands. RANGE: Genus, Silurian–Devonian; Species, Llandovery–Ludlow Series. [Syn., *Spirifer radiatus*.]

3. **Plectatrypa imbricata** (J. de C. Sowerby). (×1½.) Wenlock Series; Walsall, Staffordshire. RANGE: Genus, Silurian–Devonian; Species, Llandovery–Wenlock Series. [Syn., *Atrypa imbricata*.]

4. **Cyrtia exporrecta** (Wahlenberg). (×1.) Wenlock Series; Dudley, West Midlands. RANGE: Silurian.

5. **Howellella elegans** (Muir-Wood). (×2.) Wenlock Series; Dudley, West Midlands. RANGE: Genus, Silurian; Species, Llandovery–Ludlow Series. [Syn., *Delthyris elegans*.]

6. **Meristina obtusa** (J. Sowerby). (×¾.) Wenlock Series; Dudley, West Midlands. RANGE: Wenlock–Ludlow Series. [Syn., *Meristella tumida, Meristina tumida* (Dalman).]

Plate 20

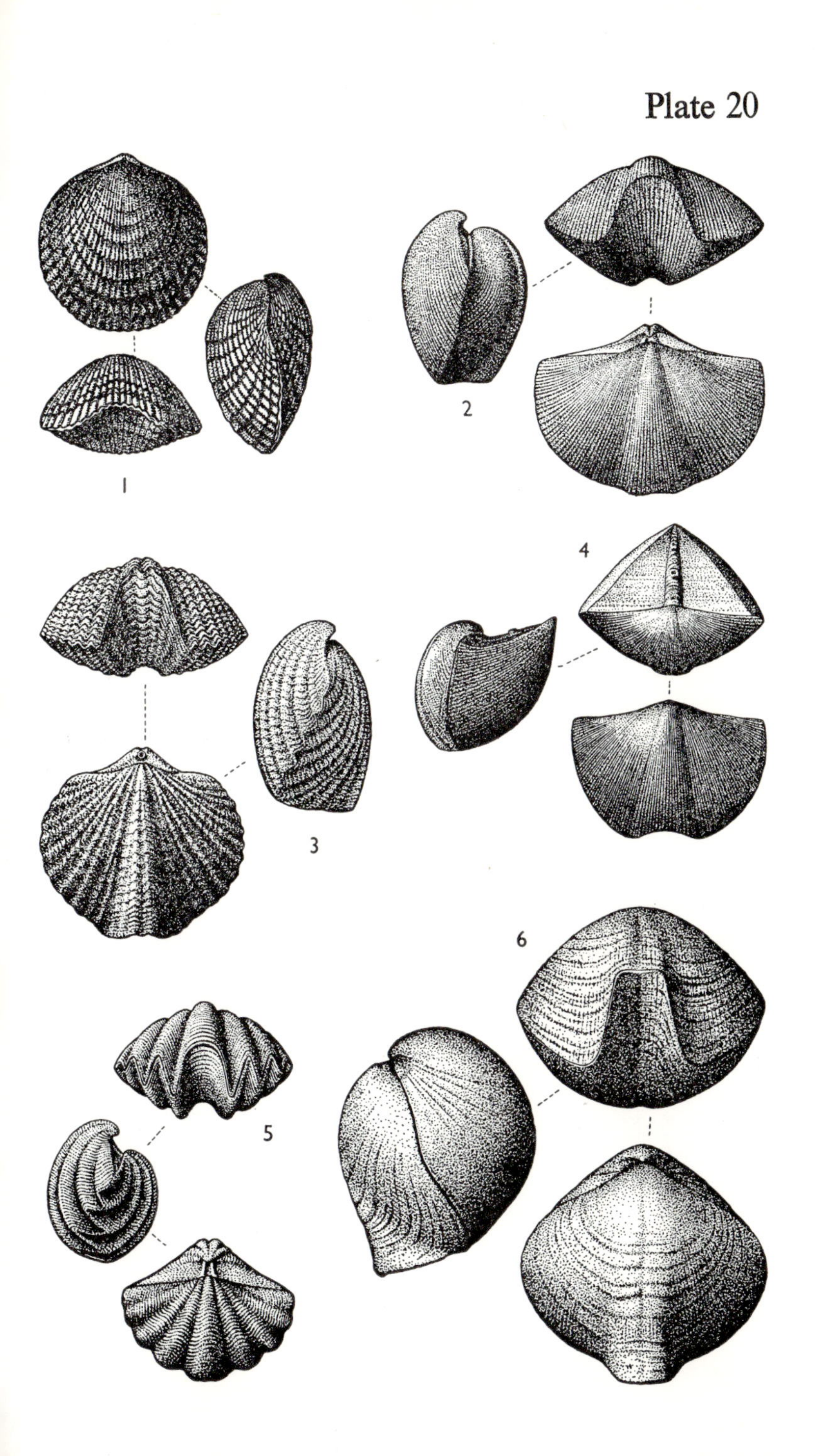

Plate 21

Silurian Echinoid (Fig. 1) and Brachiopods (Figs. 2–12)

1. **Palaeodiscus ferox** Salter. External mould of dorsal surface (×1.) Ludlow Series; Church Hill, Leintwardine, Herefordshire. RANGE: Ludlow Series.

2. **Shaleria ornatella** (Davidson). Internal mould of ventral valve. 2 (×1½), 2*a* (×5.) Ludlow Series; Whitcliffe, Ludlow, Salop. RANGE: Ludlow Series. [Syn., *Strophomena ornatella*.]

3.* **Lingula lewisi** J. de C. Sowerby. (×1.) Ludlow Series; Ledbury, Herefordshire. RANGE: Genus, Ordovician–Recent; Species, Silurian.

4. **Protochonetes ludloviensis** Muir-Wood. (×1½.) Ludlow Series; Ludlow, Salop. RANGE: Genus, Wenlock–Ludlow Series; Species, Ludlow Series. [Syn., *Chonetes striatellus* of authors.]

5. **Dayia navicula** (J. de C. Sowerby). (×2½.) Ludlow Series; Ludlow, Salop. RANGE: Genus, Silurian–Lower Devonian; Species, Wenlock–Ludlow Series. [Syn., *Terebratula navicula*.]

6, 7.* **Microsphaeridiorhynchus nucula** (J. de C. Sowerby). Internal moulds of dorsal and ventral valves. Ludlow Series. 6 (×2); Ludlow, Salop 7 (×1½), Llandigffyd, Gwent. RANGE: Species, Silurian. [Syn., *Rhynchonella nucula, Camarotoechia nucula*.]

8–10. **Salopina lunata** (J. de C. Sowerby). (×1½.) 8, ventral valve; 9,10, internal moulds of ventral and dorsal valves. Ludlow Series; near Ludlow, Salop. RANGE: Genus, Silurian–Middle Devonian (Eifelian); Species, Ludlow Series. [Syn., *Dalmanella lunata, Orthis lunata*.]

11, 12.* **Kirkidium knighti** (J. Sowerby). (×¾.) 12, transverse section. Ludlow Series; Mocktree Hill, near Leintwardine, Herefordshire. RANGE: Ludlow Series. [Syn., *Pentamerus knighti, Conchidium knighti*.]

Plate 21

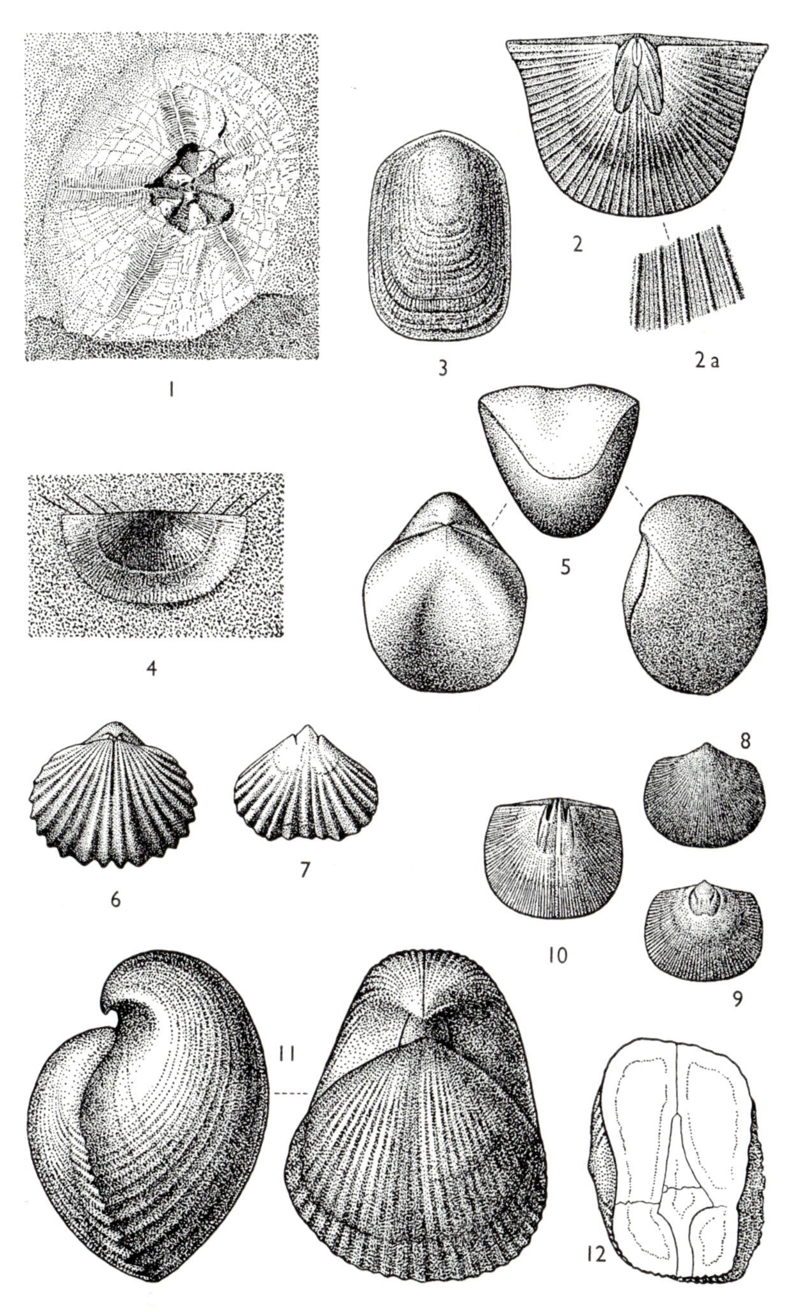

Plate 22

Silurian Crinoid (Figs. 1, 2) and Ophiuroid (Fig. 3)

1, 2.* **Crotalocrinites rugosus** (Miller). ($\times \frac{3}{4}$.) Wenlock Series; Dudley, West Midlands. RANGE: Genus, Silurian; Species, Wenlock Series. [Syn., *Crotalocrinus verrucosus* (Schlotheim).]

3.* **Lapworthura miltoni** (Salter). ($\times \frac{3}{4}$.) Ludlow Series; Leintwardine, Herefordshire. RANGE: Ordovician, Ashgill Series–Silurian, Ludlow Series. [Syn., *Lapworthura sollasi* Spencer.]

Plate 22

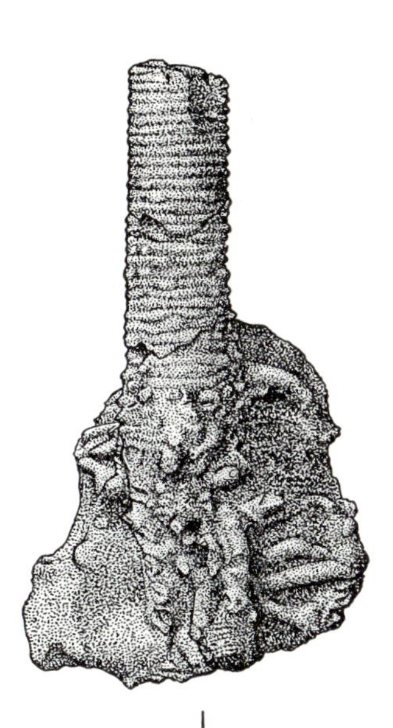

1

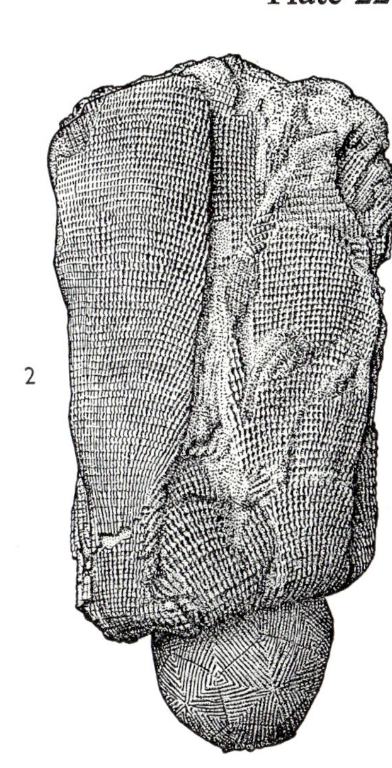

2

3

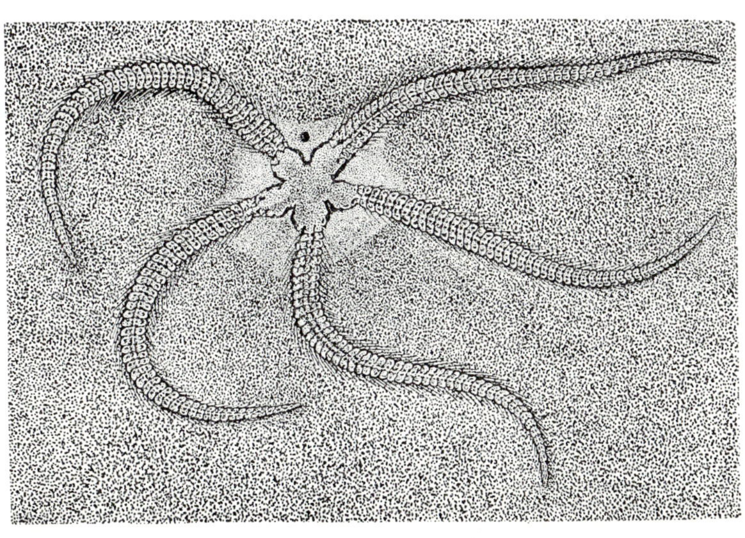

Plate 23

Silurian Crinoids (Figs. 1–3) and Calcichordate (Figs. 4, 5)

1.* **Eucalyptocrinites decorus** (Phillips). (×1.) Wenlock Series; Dudley, West Midlands. RANGE: Genus, Silurian–Devonian; Species, Wenlock Series. [Syn., *Hypanthocrinites decorus*.]

2. **Periechocrinites moniliformis** (Miller). ($\times\frac{3}{4}$.) Wenlock Series; Dudley, West Midlands. RANGE: Genus, Silurian–Carboniferous; Species, Wenlock Series. [Syn., *Periechocrinites costatus* Austin & Austin.]

3.* **Sagenocrinites expansus** (Phillips). ($\times\frac{3}{4}$.) Wenlock Series; Dudley, West Midlands. RANGE: Genus, Silurian; Species, Wenlock Series. [Syn., *Sagenocrinites giganteus* Austin & Austin.]

4, 5.* **Placocystites forbesianus** Koninck. ($\times 1\frac{1}{2}$.) Wenlock Series; Dudley, West Midlands. RANGE: Genus, Silurian–? Devonian; Species, Wenlock Series.

Plate 23

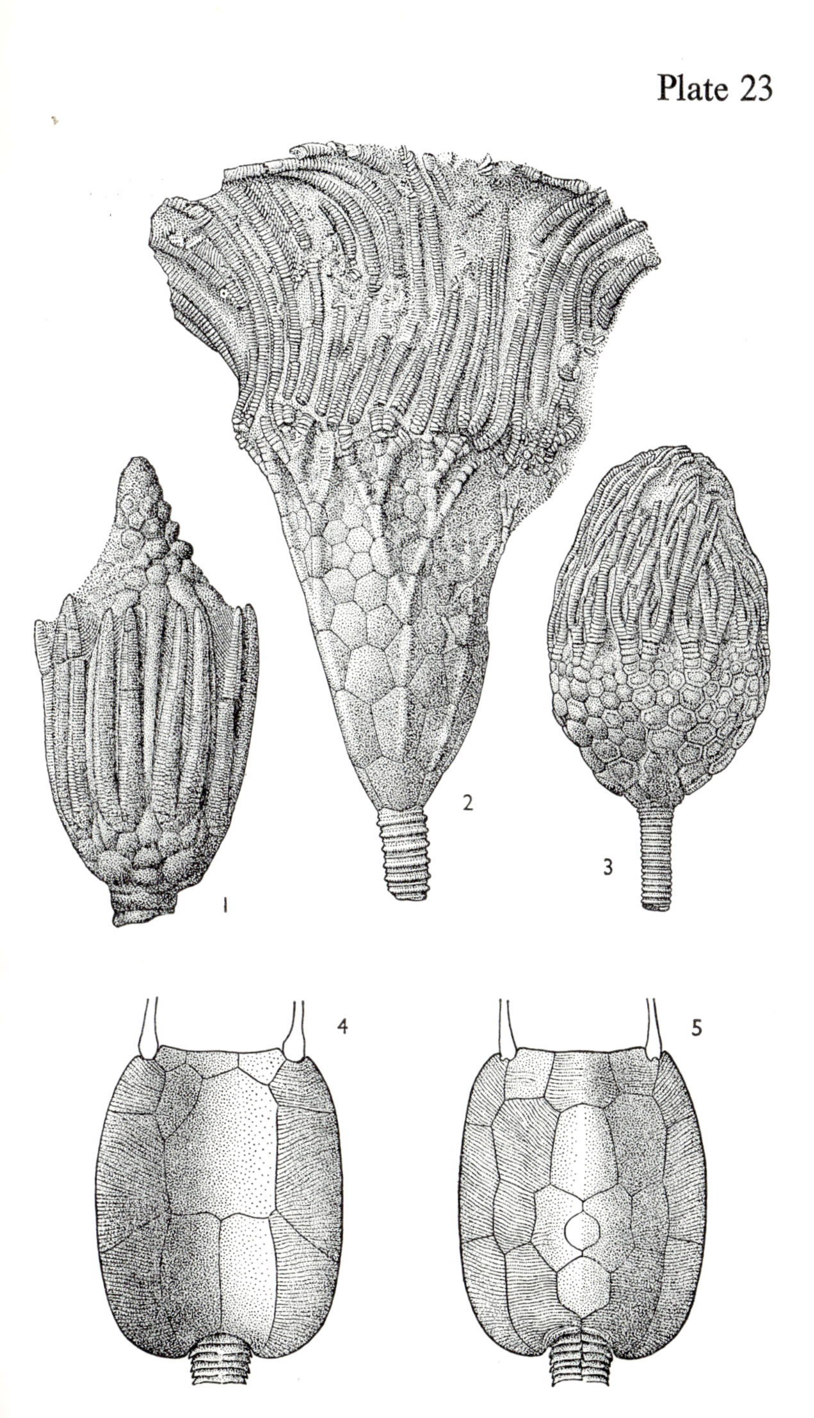

Plate 24

Silurian Cystoid (Fig. 1), Crinoid (Fig. 2) and Bivalves (Figs. 3–7)

1.* **Lepocrinites quadrifasciatus** (Pearce). (×1.) Wenlcok Series; Walsall, Staffordshire. RANGE: Genus, Silurian; Species, Wenlock Series. [Syn., *Lepadocrinus quadrifasciatus, Staurocystis quadrifasciata.*]

2.* **Gissocrinus goniodactylus** (Phillips). (×1.) Wenlock Series; Dudley, West Midlands. RANGE: Genus, Silurian–? Devonian; Species, Wenlock Series.

3.* **Pteronitella retroflexa** (Wahlenberg). (×1.) Wenlock Series; Dudley, West Midlands. RANGE: Genus, Ordovician–Silurian; Species, Silurian.

4.* **Cardiola interrupta** Broderip. (×1.) Wenlock Series; Ulverston, Cumbria. RANGE: Silurian, Wenlock–Ludlow Series.

5.* **Goniophora cymbaeformis** (J. de C. Sowerby). (×1.) Ludlow Series; Dudley, West Midlands. RANGE: Genus, Ordovician, Llandeilo Series–Silurian, Ludlow Series; Species, Ludlow Series.

6.* **Fuchsella amygdalina** (J. de C. Sowerby). (×1.) Ludlow Series; locality not known. RANGE: Ludlow Series. [Syn., *Orthonota amygdalina.*]

7. **Praeclinodonta ludensis** (Reed). (×1½.) Ludlow Series; near Ludlow, Salop. RANGE: Genus, Silurian, Wenlock–Ludlow Series; Species, Ludlow Series. [Syn., *Tancrediopsis ludensis.*]

3
1
2
4
6
5
7

Plate 25

Silurian Gastropod (Fig. 1), Monoplacophoran (Fig. 2) and Bivalves (Figs. 3, 4)

1.* **'Bembexia' lloydi** (J. de C. Sowerby). ($\times\frac{1}{2}$.) Ludlow Series; Aymestry, Herefordshire. RANGE: Silurian, Wenlock–Ludlow Series. [Syn., '*Pleurotomaria*' *lloydi*.]

2. **Tremanotus dilatatus** (J. de C. Sowerby). ($\times\frac{3}{4}$.) Wenlock Series; Dudley, West Midlands. RANGE: Genus, Upper Ordovician–Silurian; Species, Wenlock–Ludlow Series. [Syn., *Tremanotus britannicus* Newton.]

3.* **Grammysia cingulata** (Hisinger). ($\times$1.) Wenlock Series; Dudley, West Midlands. RANGE: Genus, Ordovician, Caradoc Series–Silurian, Ludlow Series; Species, Wenlock–Ludlow Series.

4.* **Palaeopecten danbyi** (M'Coy). ($\times$1.) Ludlow Series; Whitcliffe, near Ludlow, Salop. RANGE: Wenlock–Ludlow Series. [Syn., *Pterinea danbyi*.]

Plate 25

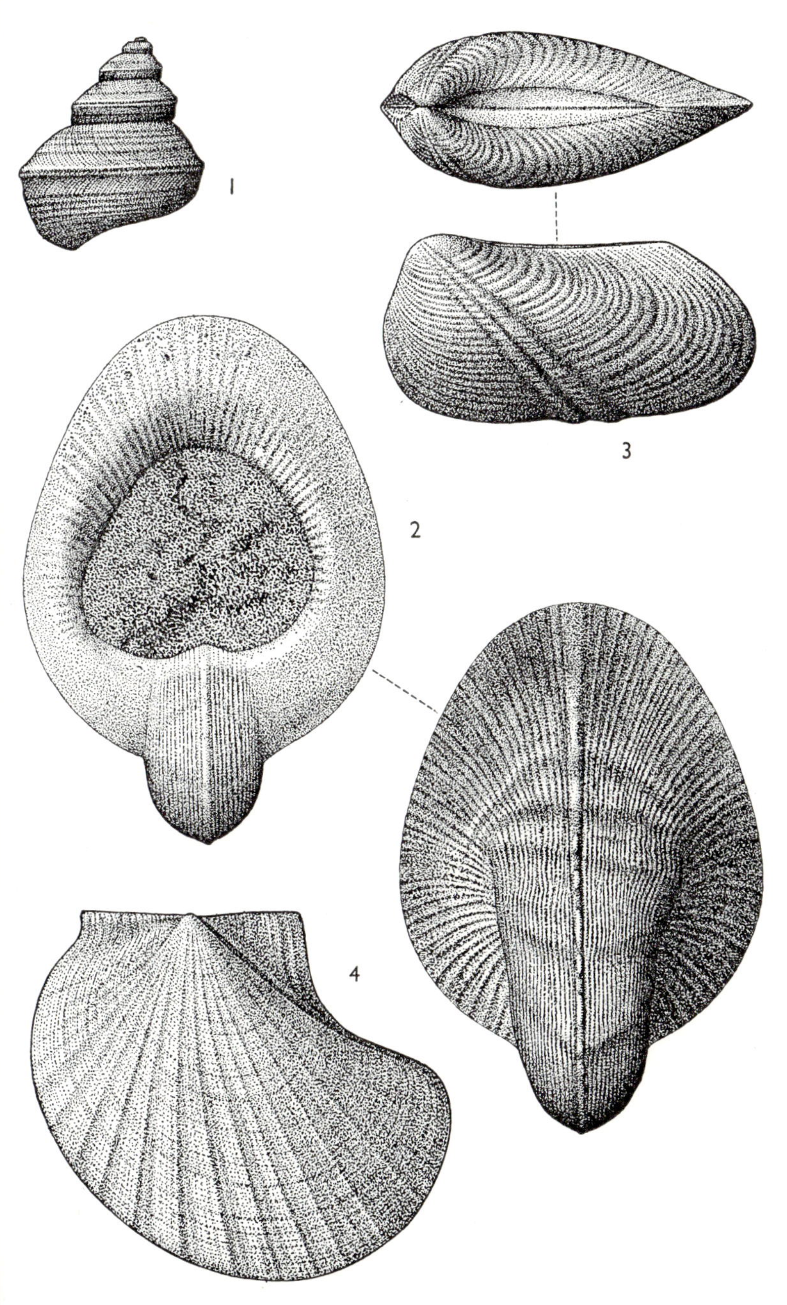

Plate 26

Silurian Gastropods

1.* **Poleumita discors** (J. Sowerby). (×1.) Wenlock Series; Benthall Edge, near Much Wenlock, Salop. RANGE: Genus, Silurian; Species, Wenlock–Ludlow Series. [Syn., *Horiostoma discors*, *Polytropina discors*.]

2, 3.* **Loxoplocus cancellatulus** (M'Coy). 2, artificial cast from external mould (×1¼), 2*a*, part of same, enlarged (×5); 3, internal mould (×1¼). Llandovery Series; Mulloch Hill, Girvan, Ayrshire. RANGE: Genus, Cambrian–Silurian; Species, Llandovery Series. [Syn., *Lophospira cancellatula*.]

4. **Loxonema gregaria** (J. de C. Sowerby). Artificial cast from external mould (×1.) Ludlow Series; Cerig y brobach, Llandovery, Dyfed. RANGE: Genus, Ordovician, Caradoc Series–Devonian; Species, Silurian, Ludlow Series. [Syn., *Holopella gregaria*.]

5. **Platyceras haliotis** (J. de C. Sowerby). (×1.) Wenlock Series; Dudley, West Midlands. RANGE: Genus, Silurian–Carboniferous; Species, Wenlock–Ludlow Series.

Plate 26

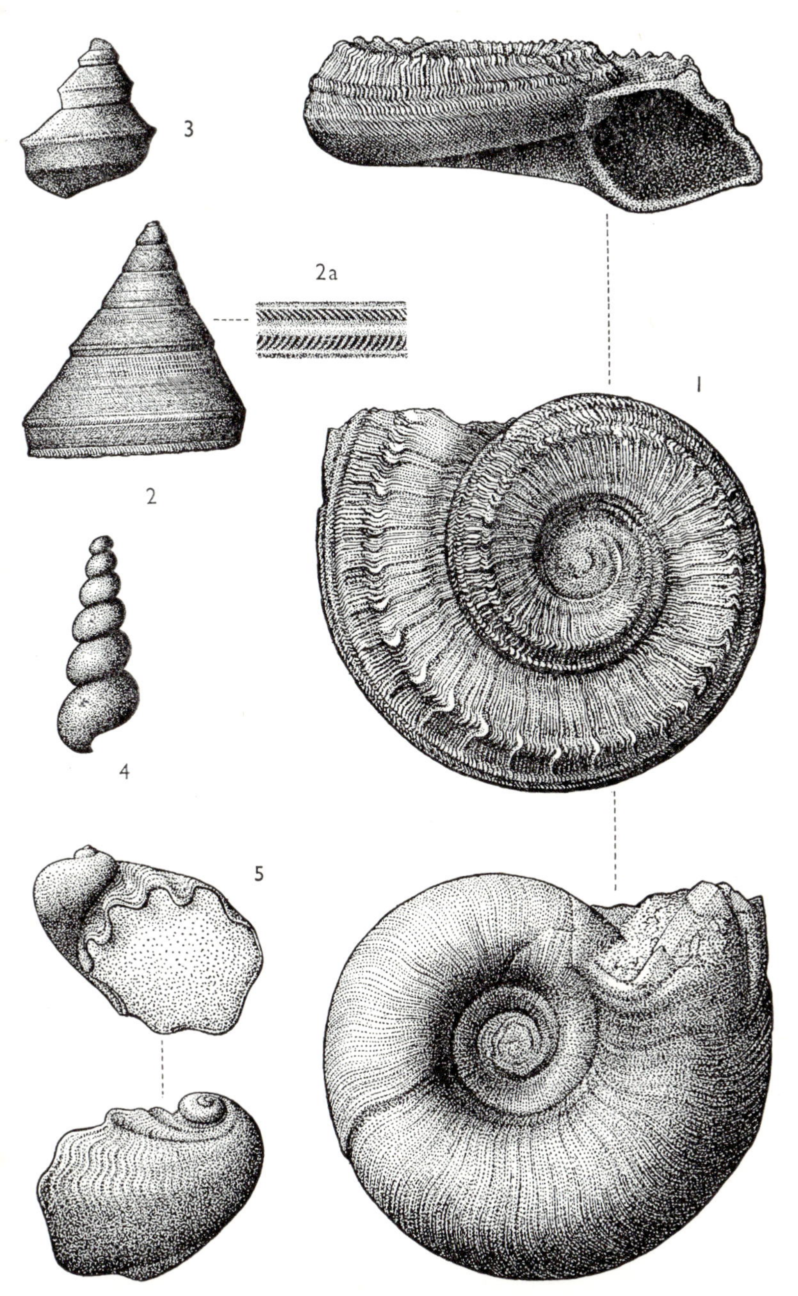

Plate 27

Silurian Gastropod (Fig. 1), Cephalopods (Figs, 2, 3) and Trilobites (Figs. 4–7)

1.* **Euomphalopterus alatus** (Wahlenberg). (×1.) Wenlock Series; near Dudley, West Midlands. RANGE: Genus, Silurian; Species, Wenlock–Ludlow Series.

2.* **Dawsonoceras annulatum** (J. Sowerby). (×¾.) Wenlock Series; Coalbrookdale, Salop. RANGE: Genus, Silurian; Species, Wenlock–Ludlow Series. [Syn., *Orthoceras annulatum*.]

3.* **Gomphoceras ellipticum** M'Coy. (×¾.) Ludlow Series; Leintwardine, Herefordshire. RANGE: Genus, Silurian; Species, Ludlow Series.

4.* **Ananaspis stokesi** (Edwards). (×1½.) Wenlock Series; Dudley, West Midlands. RANGE: Genus, Silurian–Devonian; Species, Wenlock Series. [Syn., *Phacops stokesi*.]

5. **Encrinurus onniensis** Whittard. (×3.) Llandovery Series; near Wistanstow, Salop. RANGE: Genus, Ordovician, Caradoc Series–Silurian; Species, Llandovery Series.

6, 7. **Calymene replicata** Shirley. (×2.) Llandovery Series; Newlands, Girvan, Ayrshire. RANGE: Genus, Silurian–Devonian; Species, Llandovery Series.

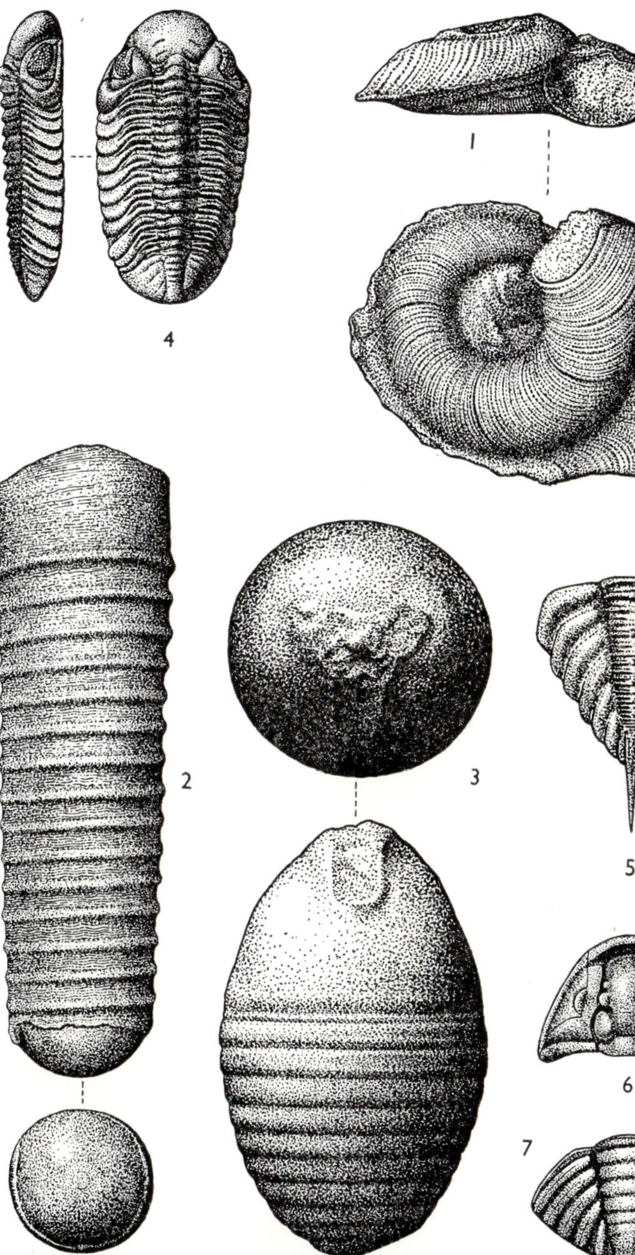
1
4
2
3
5
6
7

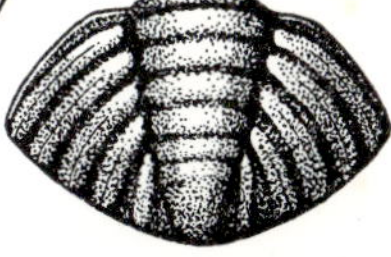

Plate 28

Silurian Trilobites

1, 2.* **Calymene blumenbachi** Brongniart. (×1.) Wenlock Series; Dudley, West Midlands. RANGE: Genus, Silurian–Devonian; Species, Wenlock Series.

3.* **Trimerus delphinocephalus** (Green). (×1.) Wenlock Series; Dudley, West Midlands. RANGE: Wenlock Series. [Syn., *Homalonotus delphinocephalus*.]

4.* **Sphaerexochus britannicus** Dean. (×1.) Wenlock Series; Malvern, Worcestershire. RANGE: Genus, Ordovician, Caradoc Series–Silurian; Species, Wenlock Series. [Syn., *Sphaerexoxochus mirus*.]

5.* **Dalmanites myops** (König). (×1¼.) Wenlock Series; Dudley, West Midlands. RANGE: Genus, Silurian–Devonian; Species, Silurian. [Syn., *Dalmanites vulgaris* (Salter), *Phacops longicaudatus* Murchison.]

6.* **Ktenoura retrospinosa** Lane. (×2.) Wenlock Series; Dudley, West Midlands. RANGE: Genus, Ordovician, Ashgill Series–Silurian; Species, Llandovery–Wenlock Series. [Syn., *Cheirurus bimucronatus*.]

7.* **Bumastus barriensis** (Murchison). (×½.) Wenlock Series; Dudley, West Midlands. RANGE: Genus, Ordovician–Silurian; Species, Wenlock Series. [Syn., *Illaenus barriensis*.]

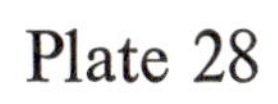
Plate 28

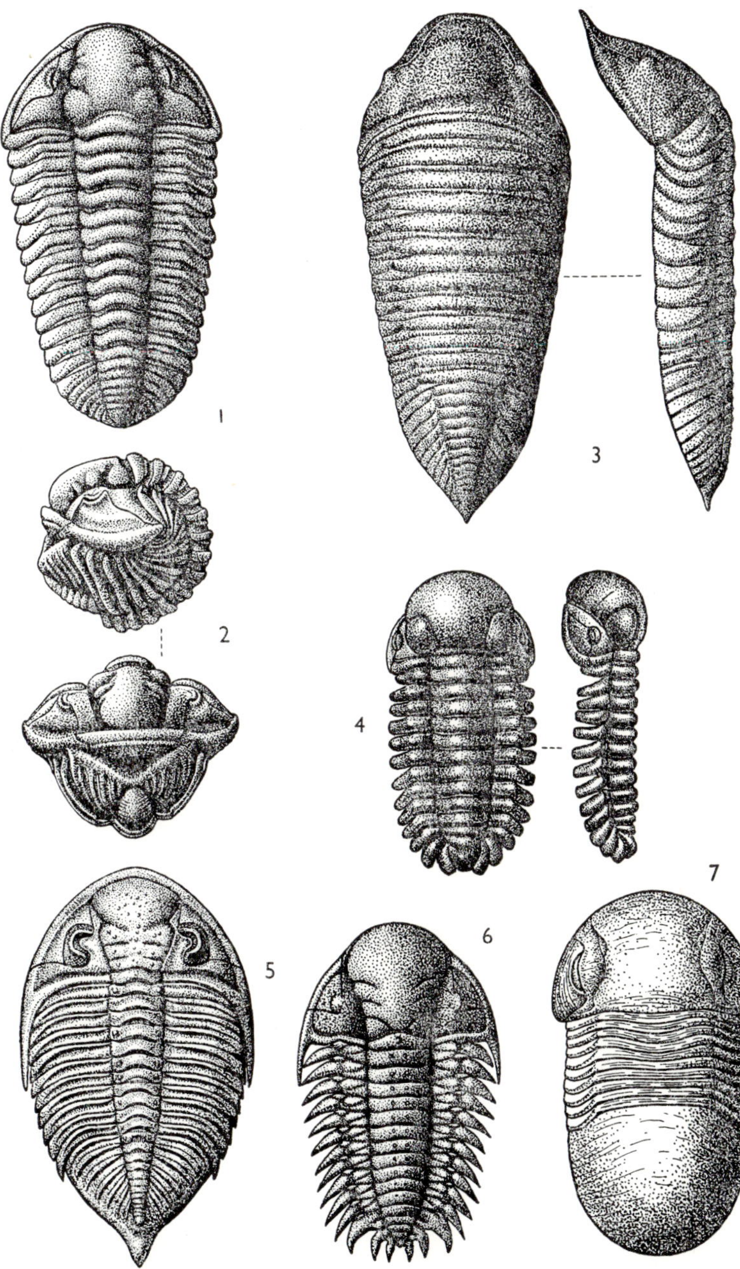

Plate 29

Silurian Ostracods (Figs. 1–3) and Trilobites (Figs. 4–10)

1, 2. **Beyrichia** cf. **kloedeni** M'Coy. (×10.) 1, male dimorph. 2, female dimorph. Wenlock Series; Coalbrookdale, near Ironbridge, Salop. RANGE: Genus, Silurian; Species, Wenlock Series.

3. **Leperditia balthica** (Hisinger). (×1.) Wenlock Series; Wren's Nest, Dudley, West Midlands. RANGE: Genus, Silurian; Species, Wenlock–Ludlow Series.

4.* **Leonaspis deflexa** (Lake). (×2.) Wenlock Series; Dudley, West Midlands. RANGE: Genus, Silurian–Devonian; Species, Wenlock Series. [Syn., *Acidaspis deflexa.*]

5.* **Deiphon barrandei** Whittard. (×2.) Wenlock Series; Dudley, West Midlands. RANGE: Genus, Silurian; Species, Wenlock Series. [Syn., *Deiphon forbesii.*]

6. **Encrinurus punctatus** (Wahlenberg). (×1.) Wenlock Series; Malvern, Worcestershire. RANGE: Genus, Ordovician, Caradoc Series–Silurian; Species, Llandovery–Wenlock Series.

7.* **Encrinurus variolaris** (Brongniart). (×1½.) Wenlock Series; Malvern, Worcestershire. RANGE: Genus, Ordovician, Caradoc Series–Silurian; Species, Wenlock Series.

8.* **Acaste downingiae** (Murchison). (×1½.) Wenlock Series; Dudley, West Midlands. RANGE: Genus, Silurian; Species, Wenlock Series. [Syn., *Phacops downingiae.*]

9, 10. **Delops obtusicaudatus** (Salter). (×1.) Ludlow Series; Coldwell Quarry, Keag Castle, near Coniston, Cumbria. RANGE: Genus, Silurian, Wenlock–Ludlow Series; Species, Ludlow Series. [Syn., *Dalmanites obtusicaudatus*, *Phacops obtusicaudatus.*]

Plate 29

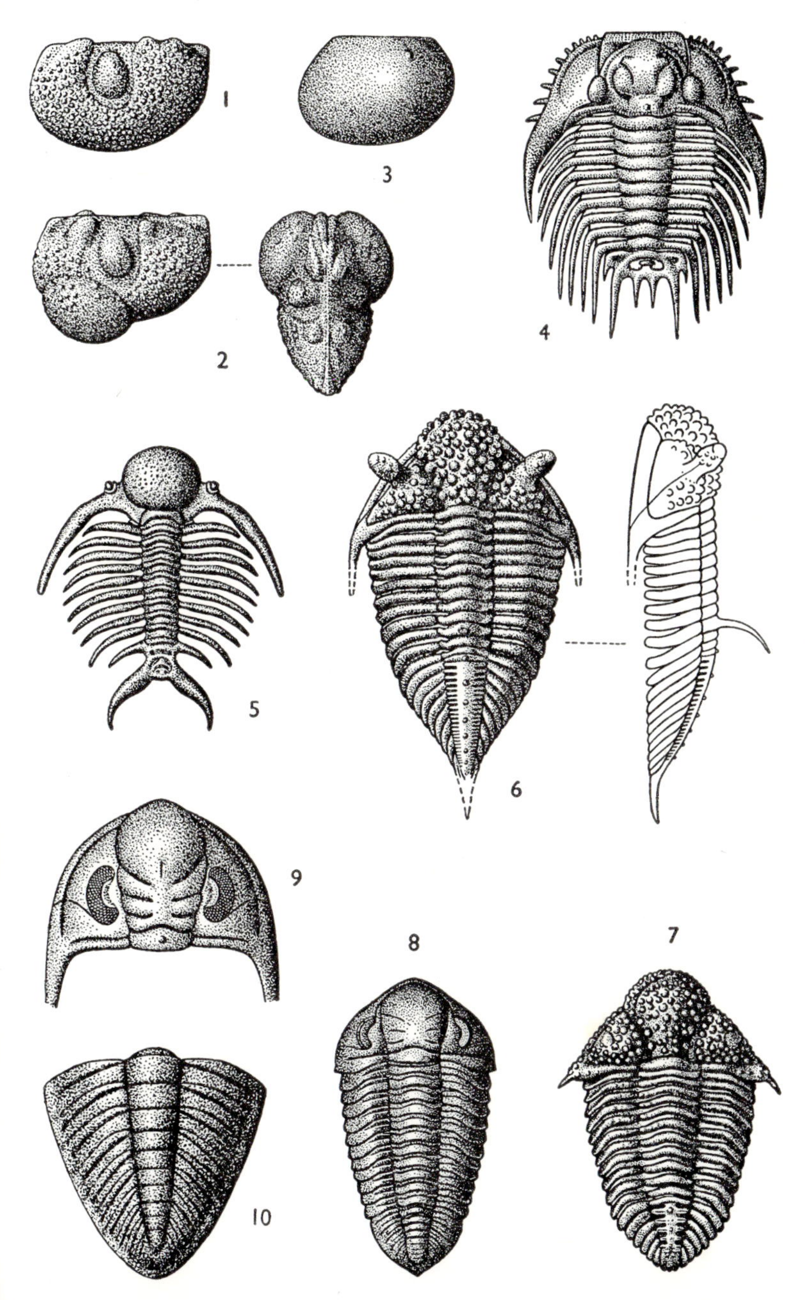

Plate 30

Silurian Graptolites (Figs. 1–9), Conodonts (Figs. 10–12), Malacostracan (Fig. 13) and Eurypterid (Fig. 14)

1.* **Monograptus sedgwickii** (Portlock). (×2.) Llandovery Series; Stockdale, near Ambleside, Cumbria. RANGE: Genus, Silurian; Species, Llandovery Series.

2.* **Monograptus priodon** (Bronn). (×2.) Llandovery Series; Grieston Quarry, south-west of Innerleithen, Peebleshire. RANGE: Genus, Silurian; Species, Llandovery–Wenlock Series.

3.* **Monograptus lobiferus** (M'Coy). (×2.) Llandovery Series; Dobbs Linn, Moffat, Dumfriesshire. RANGE: Genus, Silurian; Species, Llandovery Series.

4.* **Monograptus colonus** (Barrande). (×2.) Ludlow Series; Dudley, West Midlands. RANGE: Genus, Silurian; Species, Ludlow Series.

5. **Monograptus leintwardinensis** Hopkinson. (×2.) Ludlow Series, Leintwardine, Herefordshire. RANGE: Genus, Silurian; Species, Ludlow Series.

6. **Petalograptus minor** Elles. (×2.) Llandovery Series; Skelgill, near Ambleside, Cumbria. RANGE: Llandovery Series.

7. **Monograptus turriculatus** (Barrande). (×2.) Llandovery Series; Stockdale, near Ambleside, Cumbria. RANGE: Genus, Silurian; Species, Llandovery Series.

8. **Diplograptus modestus** (Lapworth). (×2.) Llandovery Series; Girvan, Ayrshire. RANGE: Genus, Ordovician, Llanvirn Series–Silurian, Llandovery Series; Species, Llandovery Series. [Syn., *Mesograptus modestus*.]

9. **Cyrtograptus murchisoni** Carruthers. (×2.) Wenlock Series; near Builth Wells, Powys. RANGE: Wenlock Series.

10. **Panderodus unicostatus** (Branson & Mehl). (×15.) Ludlow Series; Clungunford, Salop. RANGE: Genus, Silurian, Wenlock–Ludlow Series; Species, Ludlow Series.

11. **Ozarkodina typica** Branson & Mehl. (×20.) Ludlow Series; Clungunford, Salop. RANGE: Genus, Ordovician–Lower Carboniferous; Species, Ludlow Series.

continued opposite Plate 31

Plate 30

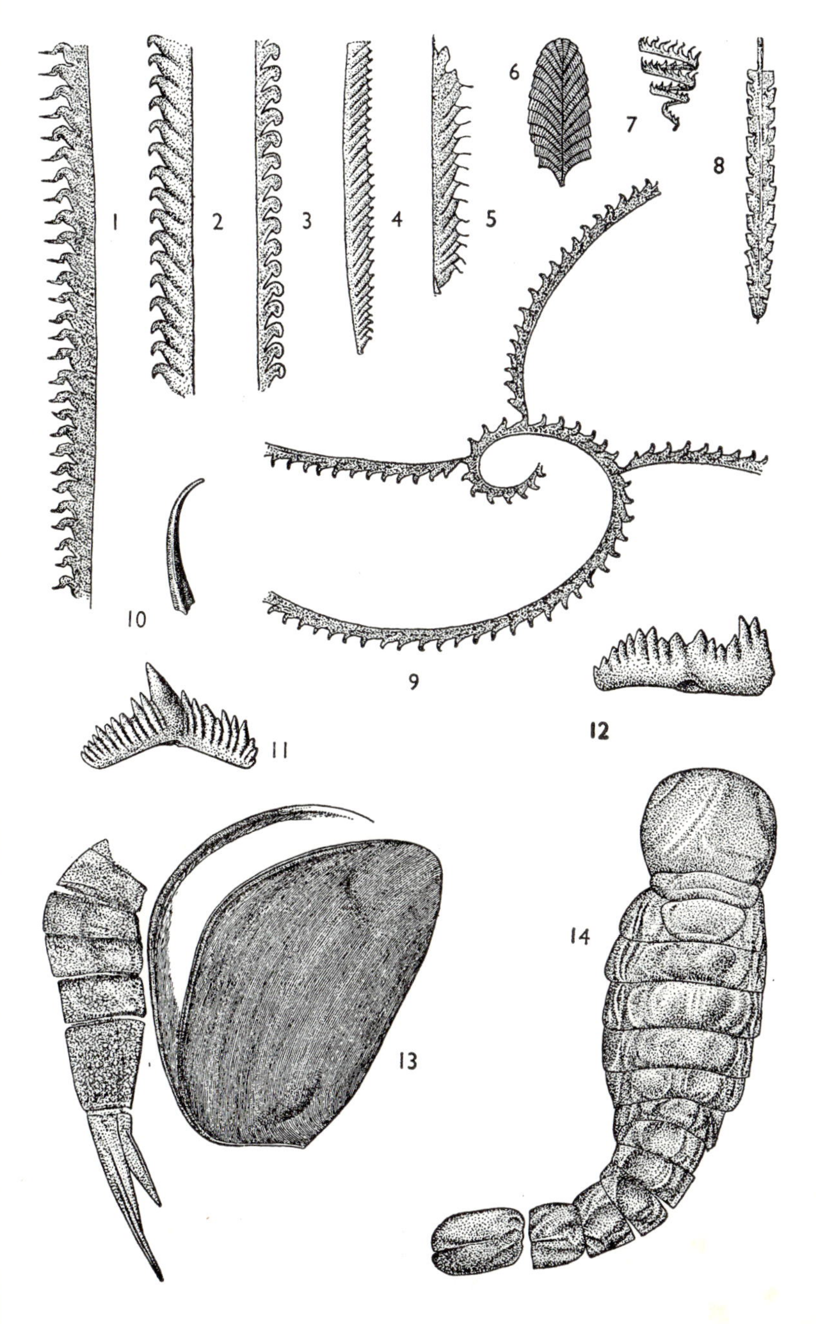

Plate 31

Devonian Plants (Figs. 1–3), Stromatoporoid (Fig. 4) and Corals (Figs. 5–9)

1. **Psilophyton princeps** Dawson. Part of stem (×¾.) Lower Old Red Sandstone, Senni Beds; Llanover, near Abergavenny, Gwent. RANGE: Lower Devonian.

2, 3. **Zosterophyllum llanoveranum** Croft & Lang. 2, axes; 3, sporangia (×1.) Lower Old Red Sandstone, Senni Beds; Llanover, near Abergavenny, Gwent. RANGE: Lower Devonian.

4. **Stromatopora huepschii** (Bargatsky). Longitudinal Section (×10.) Pebble from Middle ? Devonian, Teignmouth, Devon. RANGE: Genus, Ordovician–Permian; Species, Middle–Upper Devonian.

5.* **Favosites goldfussi** Orbigny. Transverse Section (×2.) Middle Devonian; South Devon. RANGE: Genus, Upper Ordovician–Upper Devonian (? Trias); Species, Middle–Upper Devonian.

6.* **Heliolites porosus** (Goldfuss). Transverse Section (×5.) Middle Devonian, South Devon. RANGE: Genus, Silurian–Middle Devonian; Species, Lower–Middle Devonian.

7. **Hexagonaria goldfussi** (Verneuil & Haime). Transverse Section (×3.) Upper Devonian; Babbacombe, Torquay, Devon. RANGE: Genus, Devonian; Species, Upper Devonian. [Syn., *Acervularia goldfussi.*]

8.* **Thamnopora cervicornis** (Blainville). Transverse Section (×1.) Middle Devonian; Torquay, Devon. RANGE: Genus, Silurian–Permian; Species, Middle Devonian [Syn., *Pachypora cervicornis.*]

9.* **Haplothecia pengellyi** (Edwards & Haime). Transverse section (×2.) Upper Devonian; pebble in Parson and Clerk Rocks, Dawlish, Devon. RANGE: Genus, Devonian; Species, probably Middle Devonian. [Syn., *Smithia pengellyi.*]

Plate 30 (*continued*)

12. **Spathognathodus typicus** (Branson & Mehl). (×20.) Ludlow Series; Clungunford, Salop. RANGE: Genus, Silurian–Upper Carboniferous; Species, Ludlow Series.

13.* **Ceratiocaris stygia** Salter. (×¾.) Ludlow Series, North Cumberhead, Logan Water, Lanarkshire. RANGE: Genus, Ordovician–Lower Carboniferous; Species, Ludlow Series.

14. **Erettopterus bilobus** (Salter). (×½.) Ludlow Series; Lesmahagow, Lanarkshire. RANGE: Genus, Ordovician–Devonian; Species, Ludlow Series. [Syn., *Pterygotus bilobus.*]

Plate 31

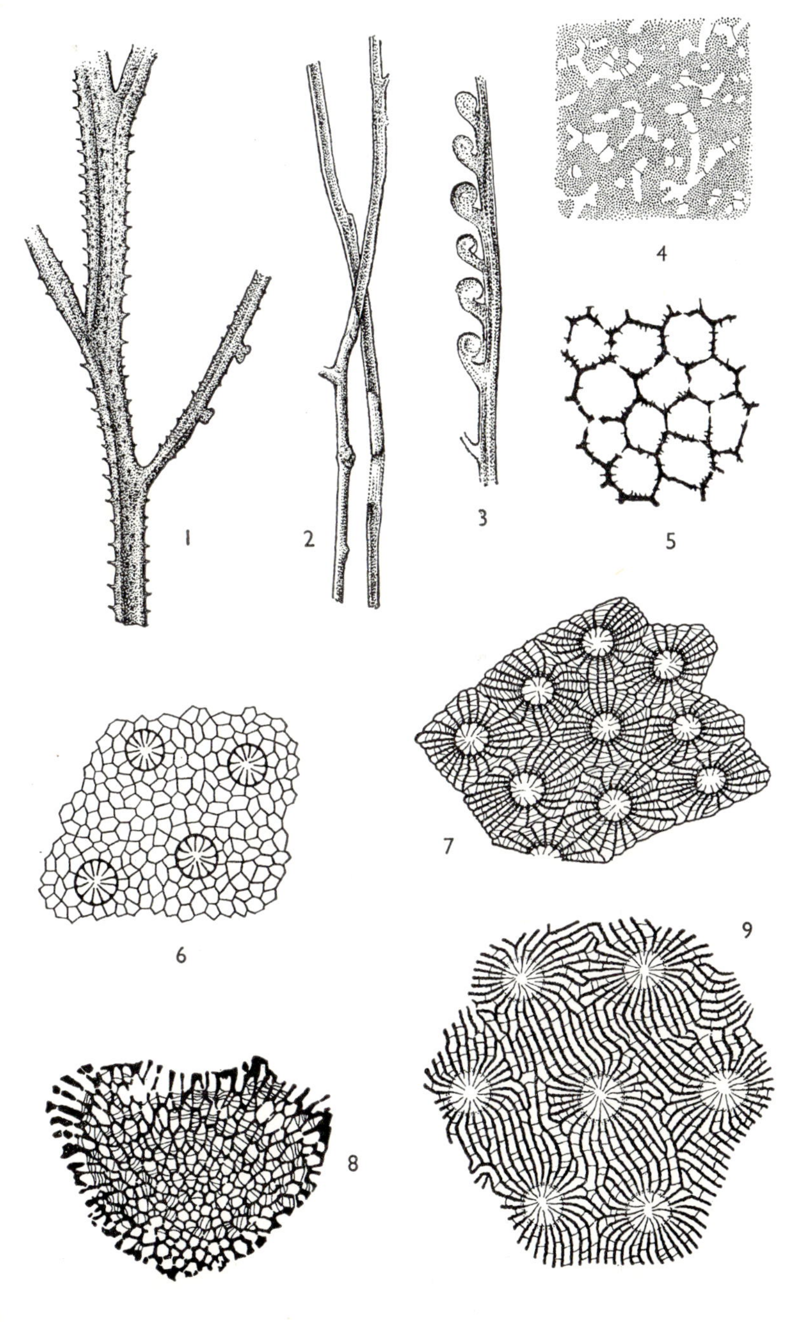

British Palaeozoic Fossils

Plate 32

Silurian and Devonian Brachiopods (Fig. 1–4), Corals (Figs. 5–7) and Crinoid (Fig. 8)

1.* **Productella fragaria** (J. de C. Sowerby). (×¾.) Devonian; Devon. RANGE: Genus, Lower Devonian–Lower Carboniferous; Species, Upper Devonian.

2. **Mesoplica praelonga** (J. de C. Sowerby). (×1.) Upper Devonian; near Tiverton, Devon. RANGE: Upper Devonian. [Syn., *Productella praelonga*.]

3. **Lingula cornea** J. de C. Sowerby. (×1.) Silurian, Downtonian Stage; Ludlow, Salop. RANGE: Genus, Ordovician–Recent; Species, Downtonian Stage.

4. **Lingula minima** J. de C. Sowerby. (×¾.) Silurian, Downtonian Stage; Ludlow, Salop. RANGE: Genus, Ordovician–Recent; Species, Downtonian Stage.

5.* **Disphyllum goldfussi** (Geinitz). Transverse section (×2). Pebble in Middle (?) Devonian; Teignmouth, Devon. RANGE: Genus, Silurian, Wenlock Series–Devonian; Species, Middle–Upper Devonian. [Syn., *Cyathophyllum caespitosum* Goldfuss in part.]

6.* **Plasmophyllum (Mesophyllum) bilaterale** (Champernowne). Transverse section (×¾.) Middle Devonian; Tuckenhay, northwest of Dartmouth, Devon. RANGE: Middle Devonian. [Syn., *Cyathophyllum? bilaterale*.]

7. **Phillipsastrea devoniensis** (Edwards & Haime). Transverse section (×1½.) Upper Devonian; Devon. RANGE: Upper Devonian. [Syn., *Pachyphyllum devoniense*.]

8.* **Hexacrinites interscapularis** (Phillips). (×1.) Middle Devonian; Wolborough, near Newton Abbot, Devon. RANGE: Genus, Devonian; Species, Middle Devonian. [Syn., *Hexacrinus melo* Austin & Austin, *Platycrinus interscapularis*.]

Plate 32

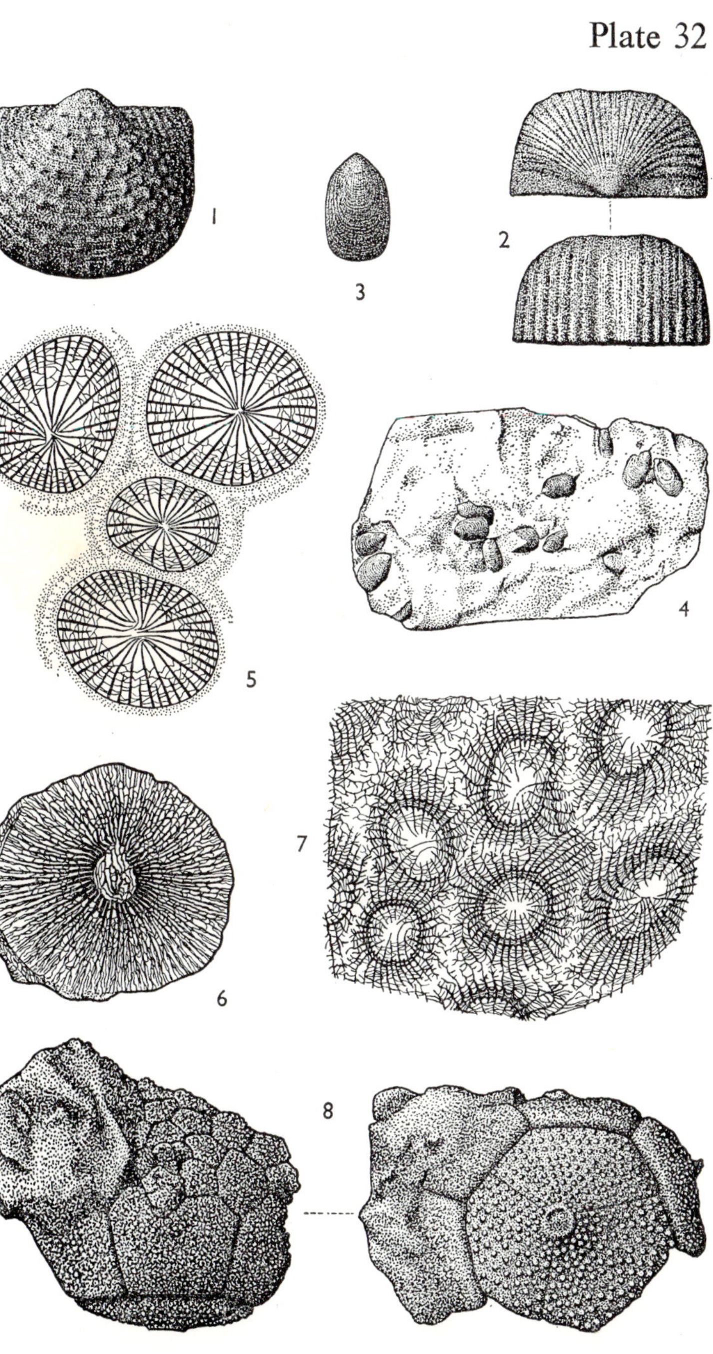

Plate 33

Devonian Brachiopods

1.* **Cyrtospirifer extensus** (J. de C. Sowerby). (×$\frac{3}{4}$.) Upper Devonian; Delabole, Launceston, Cornwall. RANGE: Genus, Middle Devonian–Lower Carboniferous; Species, Upper Devonian.

2. **Cyrtina heteroclita** Defrance. (×$1\frac{1}{2}$.) Middle–Upper Devonian; Lummaton Hill, Barton, Torquay, Devon. RANGE: Devonian.

3. **Pyramidalia simplex** (Phillips). (×$\frac{3}{4}$.) Middle Devonian; Lummaton Hill, Barton, Torquay, Devon. RANGE: Devonian. [Syn., *Cyrtia simplex*.]

4.* **Spirifer undiferus** Roemer. (×1.) Devonian; near Newton Abbot, Devon. RANGE: Genus, Devonian–Permian; Species, Middle Devonian.

5, 6. **Sieberella brevirostris** (Phillips). (×1.) Middle Devonian; Lummaton Hill, Barton, Torquay, Devon. RANGE: Genus, Silurian–Devonian; Species, Middle–Upper Devonian. [Syn., *Pentamerus brevirostris*.]

7.* **Stropheodonta nobilis** (M'Coy). (×$\frac{3}{4}$.) Middle Devonian; Lummaton Hill, Barton, Torquay, Devon. RANGE: Genus, Devonian; Species, Middle Devonian. [Syn., *Leptaena nobilis*.]

8. **Rhenorensselaeria strigiceps** (Roemer). (×$\frac{3}{4}$.) Middle Devonian. Givetian Stage; Hagginston Beach, Combe Martin Bay, Devon, RANGE: Lower Devonian. [Syn., *Rensselaeria strigiceps*.]

Plate 33

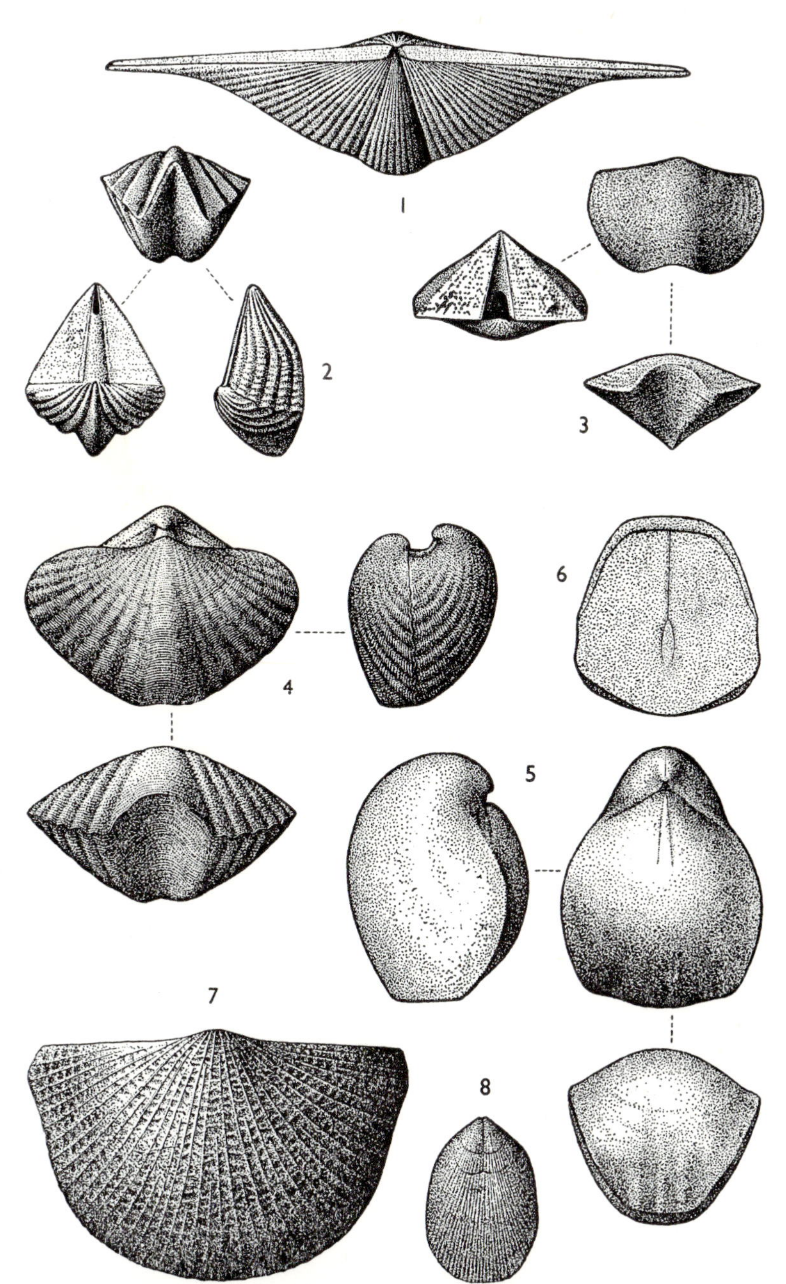

Plate 34

Devonian Brachiopods

1, 2. **Uncites gryphus** Schlotheim. (×1.) Devonian. 1, Devon (normal preservation for British specimens). 2, Paffrath near Cologne, Germany. RANGE: Devonian.

3. **Stringocephalus burtini** Defrance. (×½.) Middle Devonian; Bradley, near Newton Abbot, Devon. RANGE: Middle Devonian.

4.* **Plectatrypa aspera** (Schlotheim). (×1.) Middle Devonian; Lummaton Hill, Barton, Torquay, Devon. RANGE: Genus, Silurian–Devonian; Species, Middle Devonian. [Syn., *Atrypa aspera*.]

5.* **Hypothyridina cuboides** (J. de C. Sowerby). (×1.) Middle Devonian; Lummaton Hill, Barton, Torquay, Devon. RANGE: Devonian. [Syn., *Rhynchonella cuboides*, *Wilsonia cuboides*.]

6.* **Ladogia triloba** (J. de C. Sowerby). (×¾.) Middle Devonian; Wolborough, near Newton Abbot, Devon. RANGE: Genus, Devonian–Lower Carboniferous; Species, Middle Devonian. [Syn., *Pugnax triloba*, *Rhynchonella triloba*.]

7.* **'Camarotoechia' laticosta** (Phillips). (×¾.) Upper Devonian, Famennian Stage; Baggy Point, Croyde, near Braunton, Devon. RANGE: Species, Upper Devonian. [Syn., *Rhynchonella laticosta*.]

Plate 34

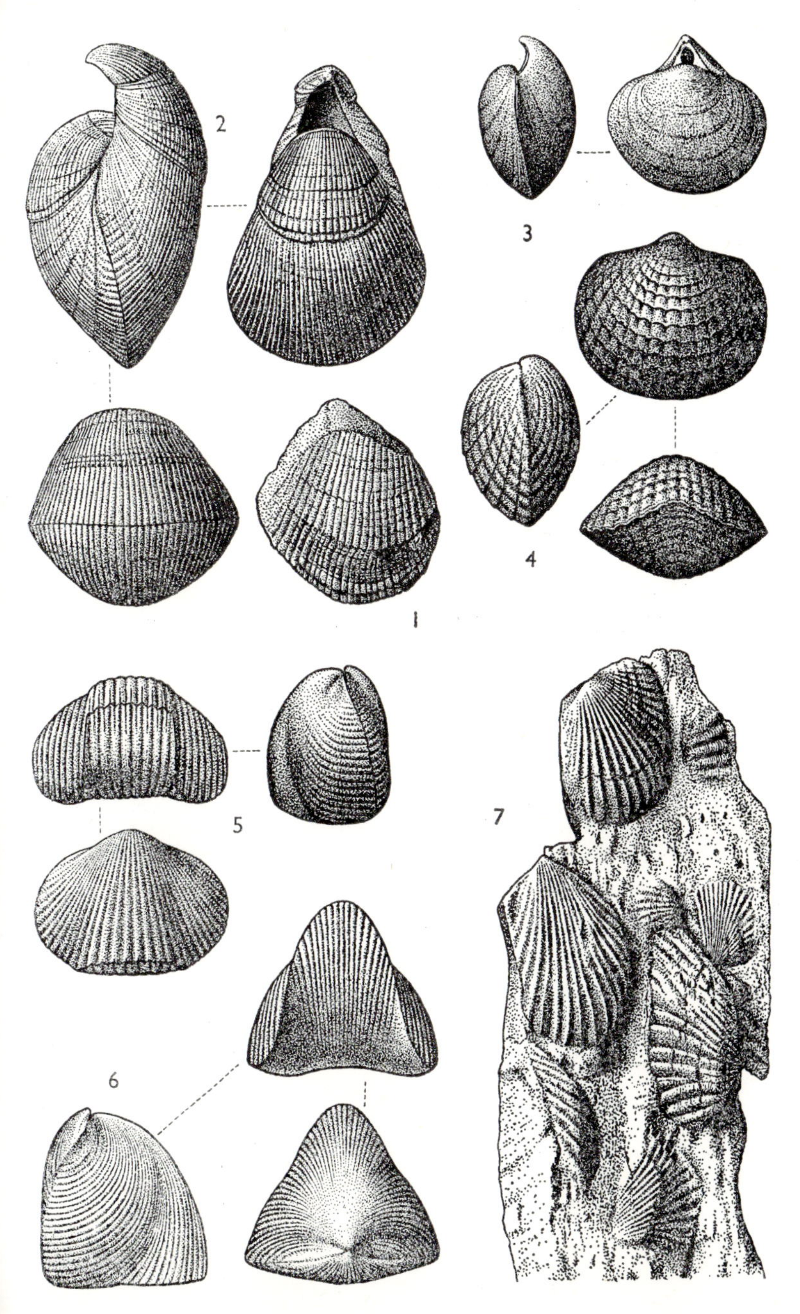

Plate 35

Devonian Bivalves (Figs. 1–3, 8), Gastropods (Figs. 4–6) and Goniatite (Fig. 7)

1.* **'Cucullaea' unilateralis** J. de C. Sowerby. (×½.) Upper Devonian; near Braunton, Devon. RANGE: Upper Devonian.

2. **Buchiola retrostriata** (Buch). (×3.) Upper Devonian; Saltern Cove, near Paignton, Devon. RANGE: Upper Devonian.

3.* **Actinopteria placida** (Whidborne). (×1½.) Middle Devonian; Lummaton Hill, Barton, Torquay, Devon. RANGE: Genus, Silurian, Wenlock Series–Lower Carboniferous; Species, Middle–Upper Devonian.

4.* **Serpulospira militaris** (Whidborne). (×1.) Middle Devonian; Newton Abbot, Devon. RANGE: Genus, Devonian–Lower Carboniferous; Species, Middle Devonian. [Syn., *Phanerotinus militaris.*]

5.* **Murchisonia bilineata** (Dechen). (×1.) Middle Devonian; Chudleigh, Devon. RANGE: Genus, Devonian–Lower Carboniferous; Species, Middle Devonian. [Syn., *Murchisonia turbinata* of authors.]

6. **Euryzone delphinuloides** (Schlotheim). (×1.) Middle Devonian; Chudleigh, Devon. RANGE: Middle–Upper Devonian.

7.* **Manticoceras intumescens** (Beyrich). (×¾.) a, septal suture. Upper Devonian; Lower Dunscombe, near Chudleigh, Devon. RANGE: Upper Devonian, *Manticoceras* Zone. [Syn., *Gephyroceras intumescens.*]

8.* **Archanodon jukesi** (Baily). (×½.) Upper Old Red Sandstone; Kiltorcan, Kilkenny, Ireland. RANGE: Upper Devonian.

Plate 35

Plate 36

Devonian Trilobites (Figs. 1–7), Conodont (Fig. 8), Goniatite (Fig. 9) and Fish (Fig. 10)

1, 2.* **Trimerocephalus mastophthalmus** (Richter). (×1.) Upper Devonian; Knowle Hill, near Newton Abbot, Devon. RANGE: Upper Devonian. [Syn., *Phacops laevis* (Münster).]

3. **Scutellum costatum** (Goldfuss). (×$\frac{3}{4}$.) Middle Devonian; Newton Abbot, Devon. RANGE: Genus, Silurian–Devonian; Species, Middle Devonian. [Syn., *Scutellum granulatum.*]

4, 5. **Dechenella setosa** (Whidborne). (×1.) Middle Devonian; Chircombe Bridge, Devon. RANGE: Middle Devonian.

6. **Phacops accipitrinus** (Phillips). (×1.) Upper Devonian; Shirwell, near Barnstaple, Devon. RANGE: Genus, Silurian–Devonian Species, Upper Devonian.

7. **Crotalocephalus pengellii** (Whidborne). (×1.) Middle Devonian; Wolborough near Newton Abbot, Devon. RANGE: Genus, Devonian; Species, Middle Devonian. [Syn., *Cheirurus pengellii.*]

8. **Icriodus** sp. (×20.) Upper Devonian; Lower Dunscombe, near Chudleigh, Devon. RANGE: Upper Devonian.

9. **Tornoceras psittacinum** (Whidborne). (×1.) Middle Devonian; Wolborough near Newton Abbot, Devon. RANGE: Middle Devonian.

10.* **Coccosteus cuspidatus** Miller. Median dorsal plate, posterior spine uppermost (×1.) Middle Old Red Sandstone; Tynet Burn, near Portgordon, Banff. (*a*) Longitudinal section (×$\frac{2}{3}$), (*b*) Transverse section (×$\frac{2}{3}$). RANGE: Genus, Middle–Upper Devonian; Species, Middle Devonian. [Syn., *Coccosteus decipiens* Agassiz.]

Plate 36

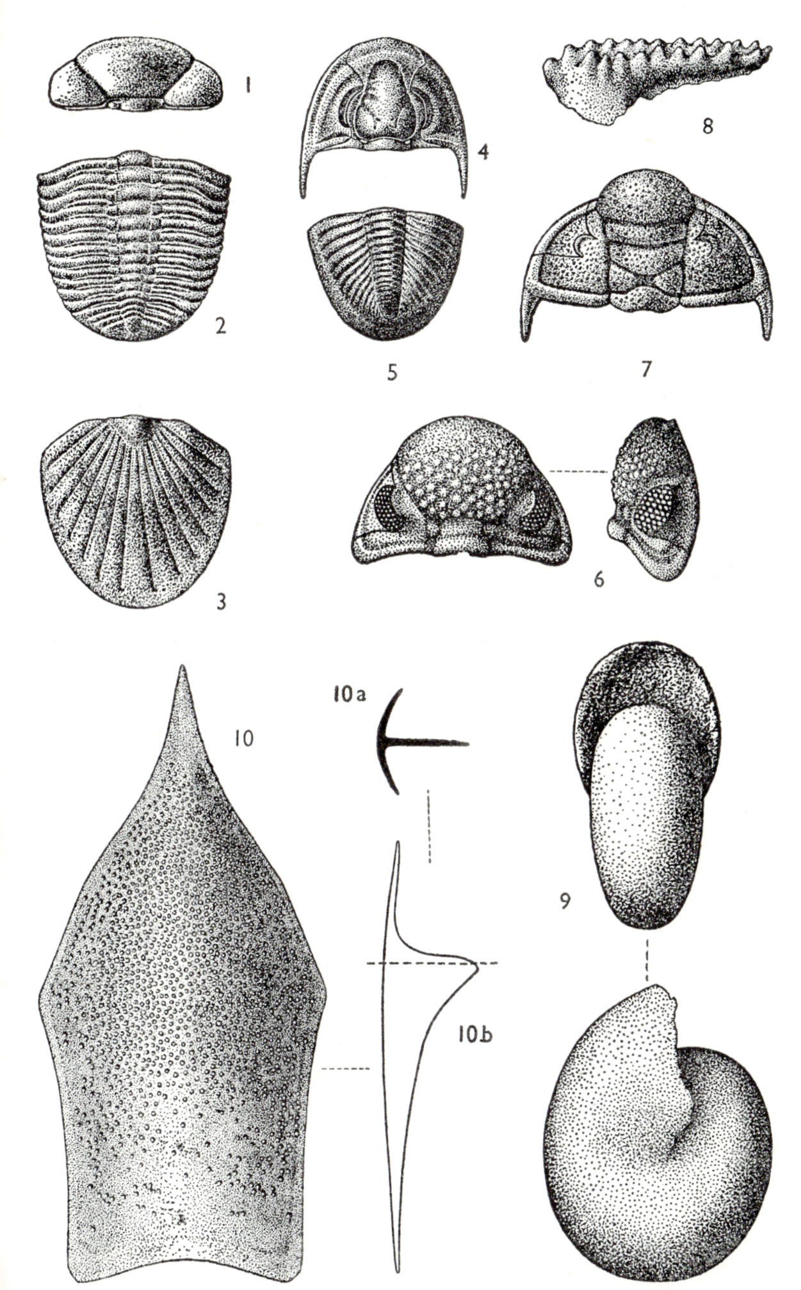

Plate 37

Devonian Agnathans (Figs. 1–3) and Fishes (Figs, 4, 5)

1. Thelodont scale. ($\times 20$.) Lower Old Red Sandstone; Hudwick Dingle, Monkhopton, near Much Wenlock, Salop. RANGE: Silurian, Llandovery Series–Devonian, Lower Old Red Sandstone.

2. **Cephalaspis lyelli** Agassiz. ($\times \frac{3}{4}$.) Lower Old Red Sandstone; Glamis, Angus. RANGE: Genus, Lower–Middle Devonian; Species, Lower Old Red Sandstone.

3. **Pteraspis rostrata** (Agassiz) subsp. **trimpleyensis** White. Dorsal Shield ($\times \frac{3}{4}$.) Lower Old Red Sandstone, Dittonian Stage; Trimpley, near Kidderminster, Worcestershire. RANGE: Genus, Lower Devonian; Species, Dittonian.

4. **Holoptychius giganteus** Agassiz. Scale ($\times \frac{3}{4}$.) Upper Old Red Sandstone; Elgin, Morayshire. RANGE: Genus and Species, Upper Devonian.

5.* **Asterolepis maxima** (Agassiz). Head Shield ($\times \frac{1}{2}$.) Upper Old Red Sandstone; King's Steps, east of Nairn (Scotland). RANGE: Genus and Species, Middle Devonian.

Plate 37

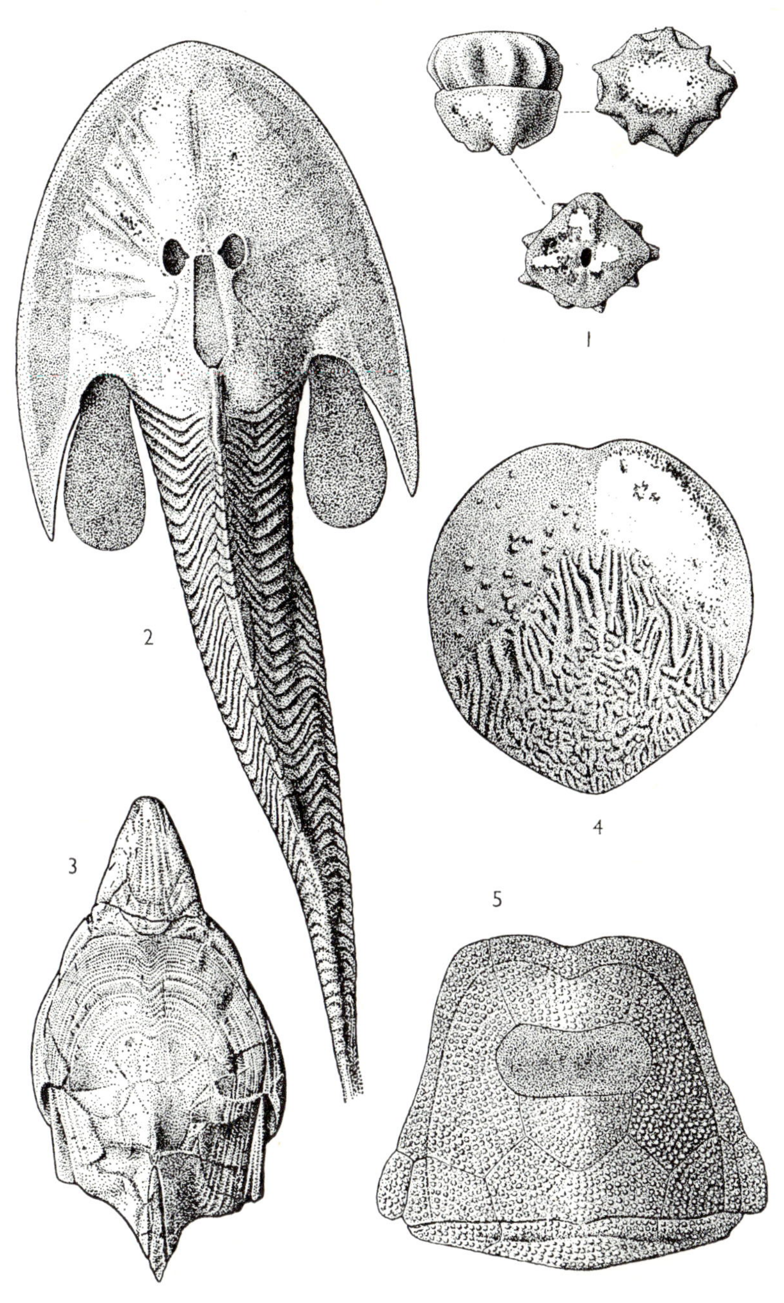
1
2
3
4
5

Plate 38

Carboniferous Plants. Articulates (Figs. 1, 2, 4, 5) and Lycopod (Fig. 3)

1. **Annularia stellata** (Schlotheim). Foliage (× ¾.) Upper Carboniferous; Clandown, Radstock, Somerset. RANGE: Genus, Carboniferous–Permian?; Species, Westphalian.

2. **Sphenophyllum emarginatum** Brongniart. Foliage (× 1.) Upper Carboniferous; Forest of Dean, Gloucestershire. RANGE: Genus, Carboniferous; Species, Westphalian.

3.* **Stigmaria ficoides** Brongniart. Part of rootstock (× ½.) Upper Carboniferous; Dudley, West Midlands. RANGE: Carboniferous.

4. **Asterophyllites equisetiformis** (Schlotheim). Foliage (× 1.) Upper Carboniferous; Radstock, Somerset. RANGE: Genus, Carboniferous; Species, Westphalian.

5.* **Calamites suckowi** Brongniart. Pith cast (× ½.) Upper Carboniferous; Gosforth, near Newcastle-on-Tyne. RANGE: Genus, Carboniferous; Species, Westphalian.

Plate 38

Plate 39

Carboniferous Plants. Pteridosperms (Figs. 1–3) and Lycopods (Figs. 4–6)

1. **Telangium affine** (Lindley & Hutton). Part of frond (×1); *a*, details of pinnules (×3.) Lower Carboniferous; West Calder, Midlothian, Scotland. RANGE: Genus, Devonian–Permian; Species, Viséan. [Syn., *Sphenopteris affinis*.]

2. **Sphenopteris alata** Brongniart. Part of frond (×1.) Upper Carboniferous; Radstock, Somerset. RANGE: Genus, Devonian–Permian; Species, Westphalian. [Syn., *Sphenopteris grandini* Goeppert.]

3. **Trigonocarpus** sp. Seed. (×1.) Upper Carboniferous; Stevenston, Ayrshire. RANGE: Upper Carboniferous.

4. **Sigillaria mamillaris** Brongniart. Part of stem (×½.) Upper Carboniferous; Darton, near Barnsley, South Yorkshire. RANGE: Genus, Carboniferous–Permian; Species, Westphalian.

5.* **Lepidodendron aculeatum** Sternberg. Part of stem or large branch (×¾.) Upper Carboniferous; Sunderland. RANGE: Genus, Carboniferous; Species, Namurian–Westphalian.

6. **Lepidodendron sternbergi** Brongniart. Leafy branch (×1.) Upper Carboniferous; Coseley, Bilston, West Midlands. RANGE: Genus, Carboniferous; Species, Westphalian.

Plate 39

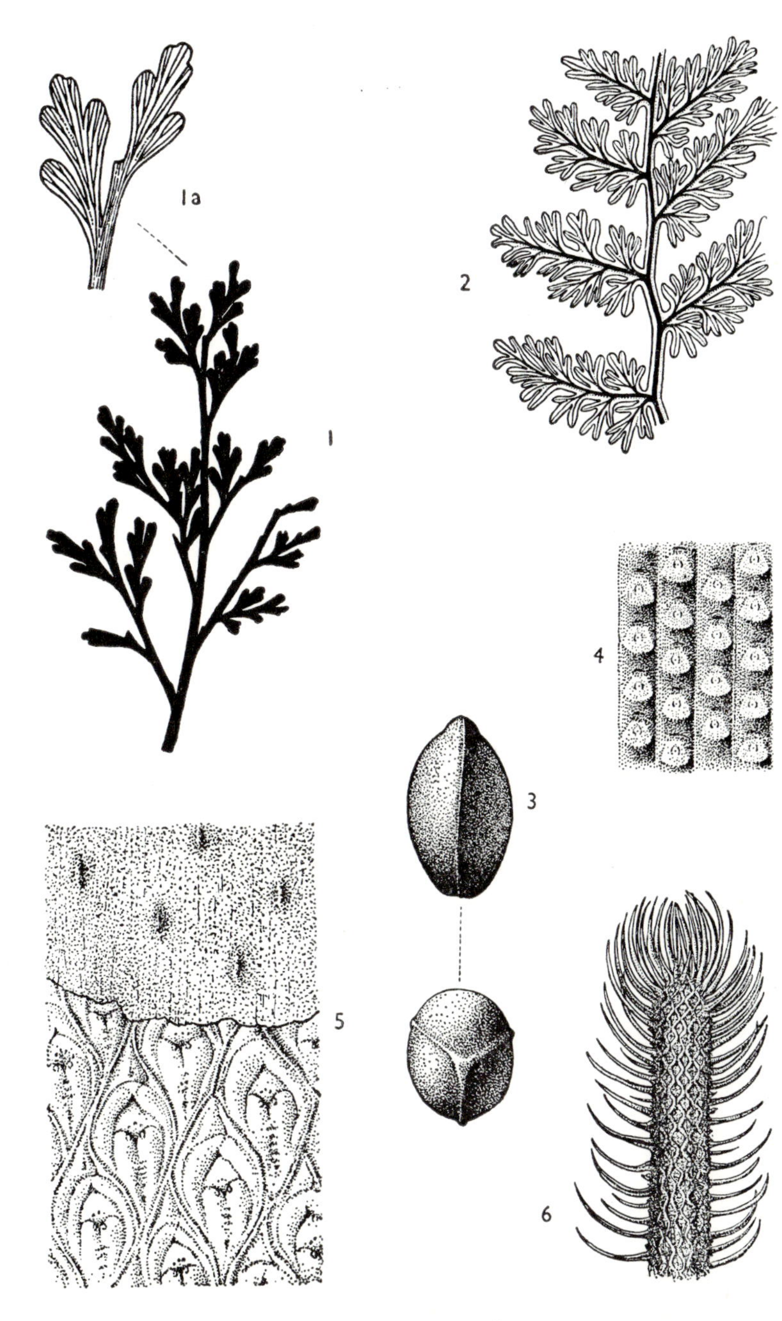

Plate 40

Carboniferous Plants. Fern (Fig. 1) and Pteridosperms (Figs. 2–5)

1. **Pecopteris polymorpha** Brongniart. Part of frond (×1); *a*, detail of sterile pinnules (×3.) Upper Carboniferous; Radstock, Somerset. RANGE: Genus, Upper Carboniferous; Species, Westphalian. [Syn., *Acitheca polymorpha*.]

2, 3.* **Mariopteris nervosa** (Brongniart). Part of frond (×1.) Upper Carboniferous, Coal Measures; Netherton, near Dudley, West Midlands. RANGE: Genus, Upper Carboniferous; Species, Westphalian.

4. **Rhodea tenuis** Gothan. Part of frond (×1.) Lower Carboniferous; Gwaenysgor, near Rhyl, Clwyd. RANGE: Genus, Carboniferous; Species, Viséan.

5.* **Alethopteris serli** Brongniart. Part of frond (×½); *a*, detail of pinnules (×3.) Upper Carboniferous; Newcastle-on-Tyne. RANGE: Genus, Carboniferous; Species, Westphalian.

1
1a
2
3
4
5
5a

Plate 41

Carboniferous Sponge (Fig. 1) and Plants, Pteridosperms (Figs. 3, 4, ? Fig. 2) and Cordaite (Fig. 5)

1. **Hyalostelia smithi** Young & Young. (×1.) Lower Carboniferous; near Richmond, North Yorkshire. RANGE: Lower Carboniferous.

2. **Rhacopteris petiolata** Goeppert. Part of frond (×1.) Lower Carboniferous; Teilia, near Prestatyn, Clwyd. RANGE: Genus, Carboniferous; Species, Viséan.

3.* **Neuropteris gigantea** Sternberg. Part of frond (×½); *a*, detail of pinnule (×2½.) Upper Carboniferous; Coseley, near Bilston, West Midlands. RANGE: Genus, Carboniferous; Species, Westphalian.

4. **Cyclopteris trichomanoides** Brongniart. Pinnule (×¾.) Upper Carboniferous; Coalbrookdale, near Ironbridge, Salop. RANGE: Genus, Upper Carboniferous; Species, Westphalian.

5.* **Cordaites angulosostriatus** Grand'Eury. Part of leaf (×¾.) Upper Carboniferous; Camerton, Somerset. RANGE: Genus, Carboniferous–Permian; Species, Westphalian.

Plate 41

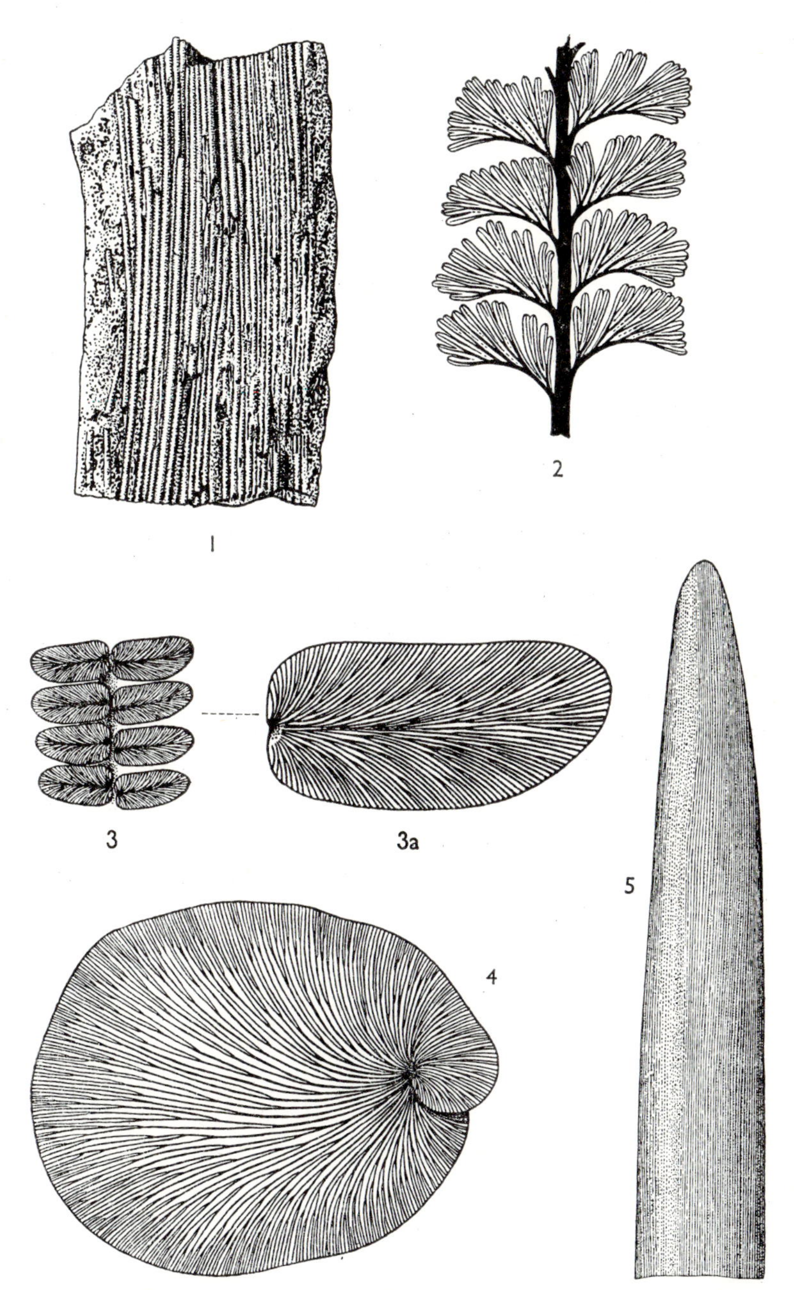

Plate 42

Carboniferous Foraminifera

1, 2. **Archaediscus karreri** Brady. (×30.) Lower Carboniferous; Brockley, Lanarkshire. RANGE: Viséan.

3, 4. **Tetrataxis conica** Ehrenberg. 3 (×25.) Lower Carboniferous; Colster Clough, near Elsdon, Northumberland. 4, Section (×40). Brockley, Lanarkshire. RANGE: Viséan–Namurian. [Syn., *Valvulina palaeotrochus* Ehrenberg.]

5. **Endothyranopsis crassa** (Brady). (×15.) Carboniferous Limestone; Great Ormes Head, Llandudno, Gwynedd. RANGE: Viséan. [Syn., *Endothyra crassa.*]

6. **Plectogyra bradyi** (Mikhailov). Section (×40.) Viséan; Locality unknown. RANGE: Viséan. [Syn., *Endothyra bowmani* of authors.]

7. **Stacheoides polytremoides** (Brady). (×10.) Viséan; Hairmyres, near East Kilbride, Lanarkshire. RANGE: Viséan–Namurian. [Syn., *Stacheia polytremoides.*]

8, 9. **Stacheia pupoides** Brady. (×50.) Namurian. 8, Downholme, near Richmond, North Yorkshire. 9, Section. Fourstones, Northumberland. RANGE: Namurian.

10, 11. **Howchina bradyana** (Howchin). Lower Carboniferous, Viséan; 10 (×60.) Tipalt, Northumberland. 11, Section (×85.) Aldfield, near Ripon, Yorkshire. RANGE: Middle–Upper Viséan.

12, 13. **Climacammina antiqua** Brady. Lower Carboniferous. 12 (×25.) 13, Section (×20.) Brockley, Lanarkshire. RANGE: Upper Viséan.

14. **Lugtonia concinna** (Brady). (×50.) Namurian; Hurst, Reeth, near Richmond, North Yorkshire. RANGE: Upper Viséan–Namurian. [Syn., *Nodosinella concinna.*]

Plate 42

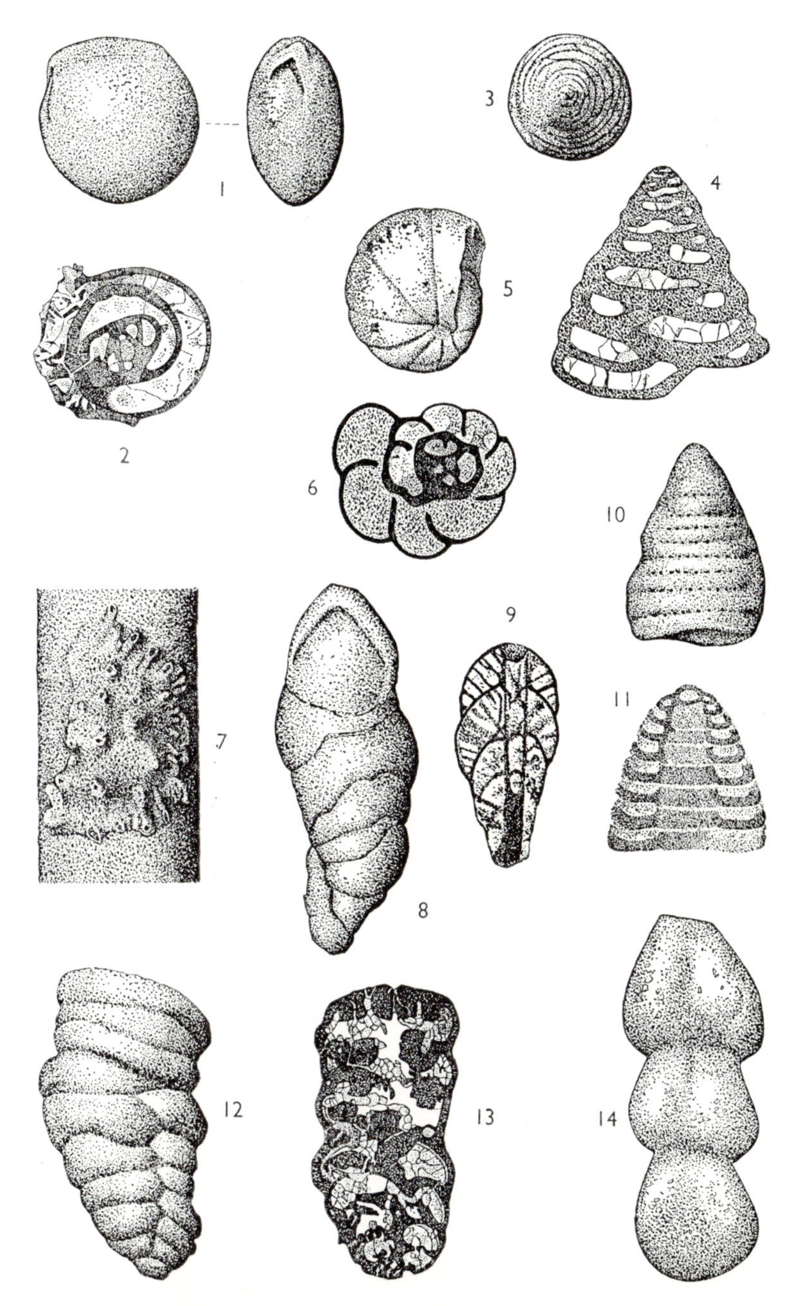

Plate 43

Carboniferous Corals

1.* **Dibunophyllum bipartitum** (M'Coy). (×1.) Viséan; Avon Gorge, Avon. RANGE: Genus, Carboniferous; Species, Viséan–Namurian.

2, 3.* **Lithostrotion junceum** (Fleming). 2, Section (×3.) Locality unknown. 3 (×1.) Viséan; Birtley, south-east of Bellingham, Northumberland. RANGE: Genus, Carboniferous; Species, Viséan–Namurian.

4, 5.* **Lithostrotion portlocki** (Bronn). Viséan. 4 (×1.) Locality unknown. 5 (×3.) Avon Gorge, Avon. RANGE: Genus, Carboniferous; Species, Viséan.

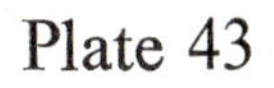

Plate 43

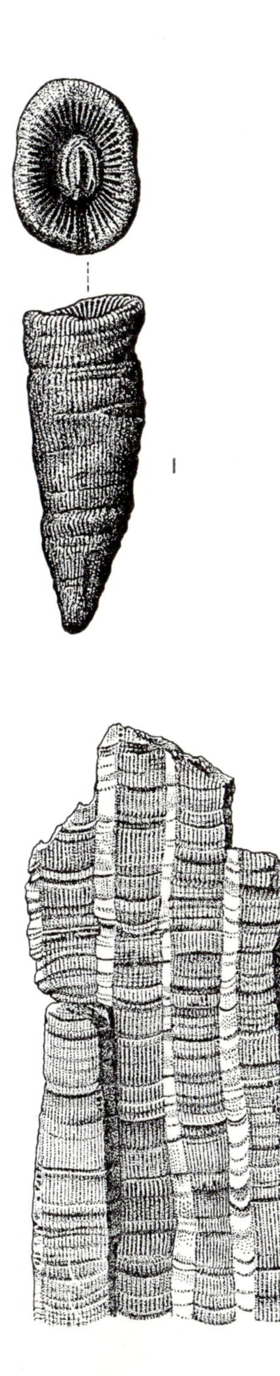

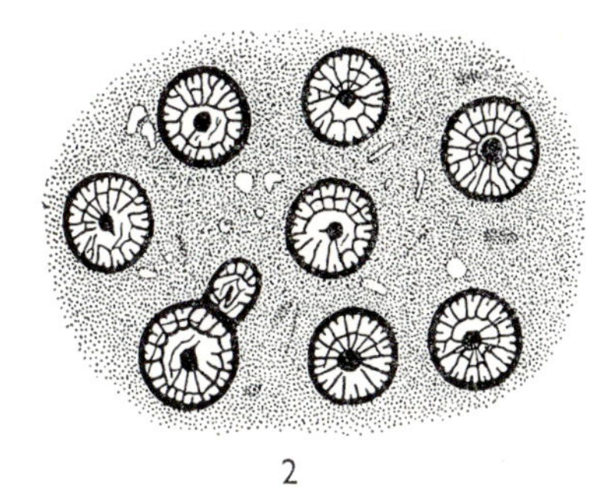

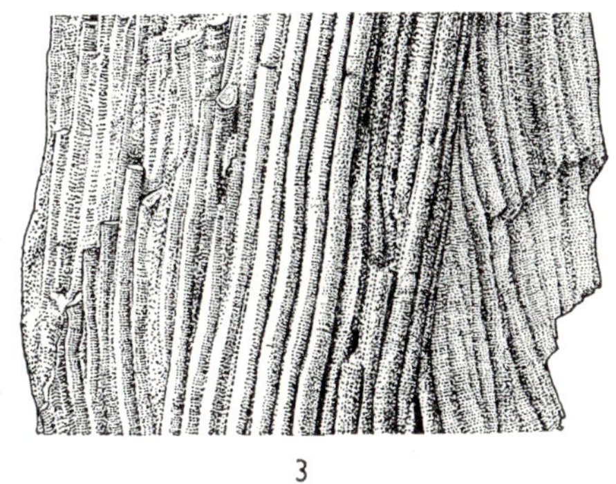

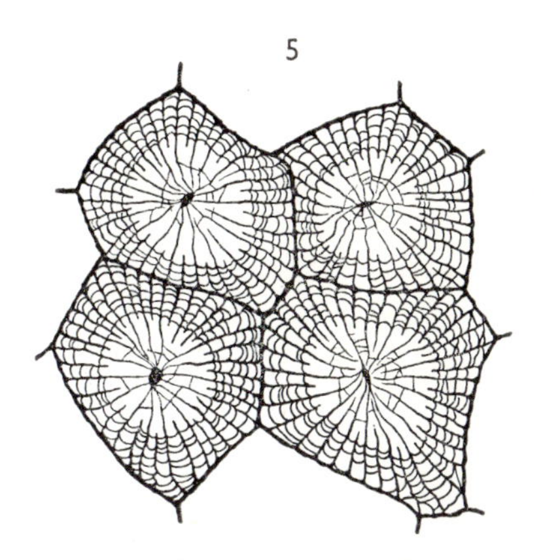

Plate 44

Carboniferous Corals

1. **Amplexus coralloides** J. Sowerby. ($\times \frac{3}{4}$.) Viséan; Derbyshire. RANGE: Genus, Lower Carboniferous; Viséan.

2. **Aulophyllum fungites** (Fleming). ($\times$ 1.) Viséan; Oswestry, Salop. RANGE: Genus, Lower Carboniferous; Species, Viséan.

3. **Amplexizaphrentis enniskilleni** (Edwards & Haime) var. **derbiensis** Lewis. Section ($\times$ 2.) Viséan; Matlock, Derbyshire. RANGE: Genus, Carboniferous; Species, Viséan. [Syn., *Zaphrentis enniskilleni.*]

4.* **Lonsdaleia floriformis** (Fleming). ($\times 1\frac{1}{2}$.) Carboniferous; Coalbrookdale, Salop. RANGE: Genus, Carboniferous; Species, Viséan–Namurian.

5.* **Palaeosmilia regium** (Phillips). ($\times$ 1.) Viséan; Clifton, Bristol. RANGE: Genus, Carboniferous; Species, Viséan–Namurian.

6, 7.* **Palaeosmilia murchisoni** Edwards & Haime. Viséan. 6, Section ($\times$ 1.) Narrowdale Grange, Alstonfield, Staffordshire. 7 ($\times \frac{3}{4}$.) Clifton, Bristol. RANGE: Genus, Lower Carboniferous; Species, Viséan.

Plate 44

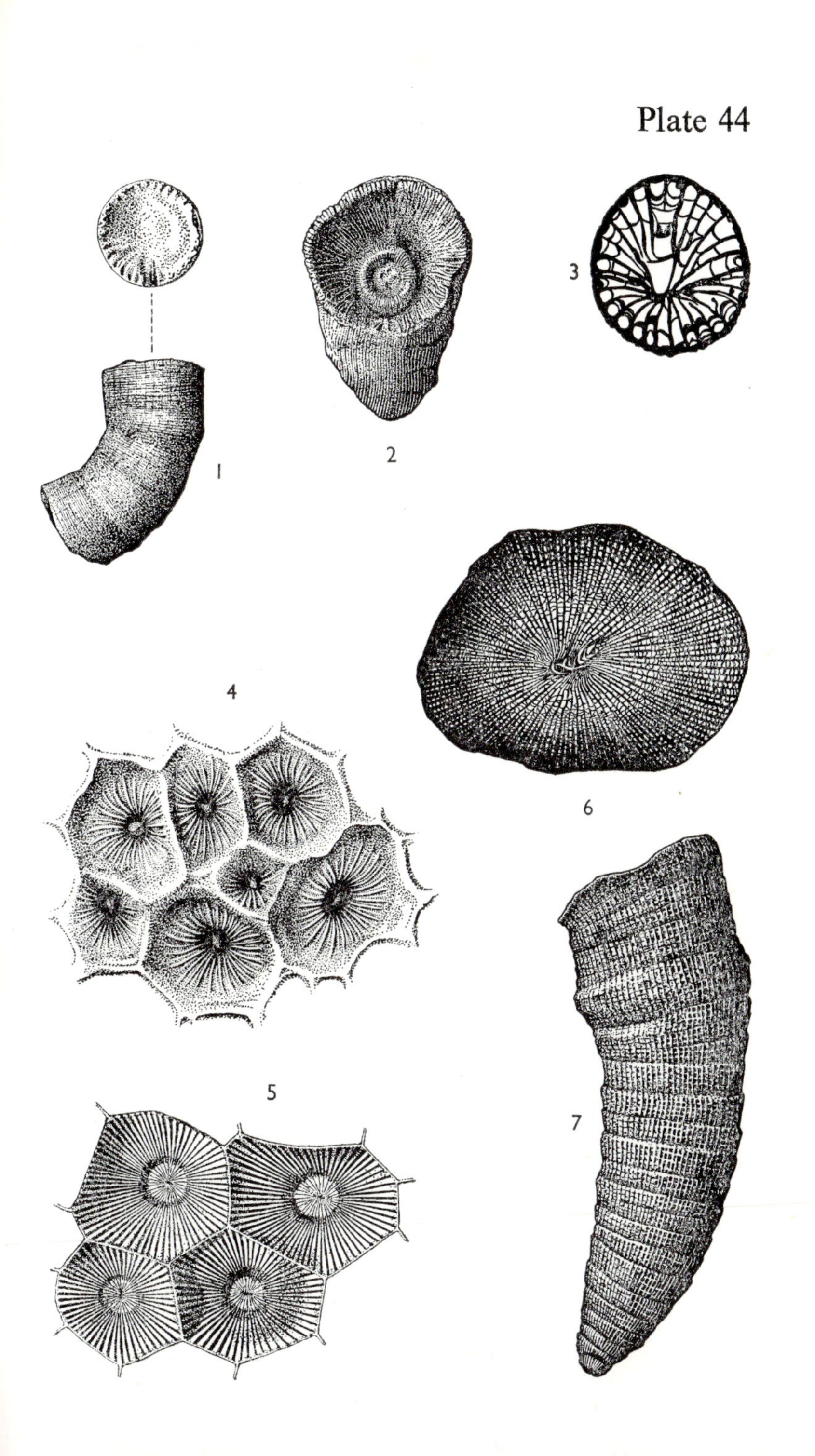

Plate 45

Carboniferous Bryozoan (Fig. 1), Corals (Figs. 2–4) and Conulata (Fig. 5)

1.* **Fenestella plebeia** M'Coy. (×2); *a* (×3.) Lower Carboniferous; Ravenstonedale, Cumbria. RANGE: Genus, Ordovician–Permian; Species, Carboniferous, Tournaisian–Viséan.

2.* **Caninia cylindrica** (Scouler). (×¾); *a*, polished surface. Lower Carboniferous; south-western England. RANGE: Lower Carboniferous. [Syn., *Caninia gigantea*.]

3.* **Syringopora geniculata** Phillips. (×1½.) Lower Carboniferous; near Limerick, Ireland. RANGE: Genus, Silurian–Carboniferous; Species, Viséan.

4.* **Michelinia tenuisepta** (Phillips). (×¾.) Tournaisian; St Thomas' Head, north of Weston-super-Mare, Avon. RANGE: Genus, Upper Devonian–Permian; Species, Tournaisian–Viséan.

5.* **Paraconularia quadrisulcata** (J. Sowerby). (×1.) Upper Carboniferous, Coal Measures; Coalbrookdale, near Ironbridge, Salop. RANGE: Genus, Ordovician–Carboniferous; Species, Carboniferous.

Plate 45

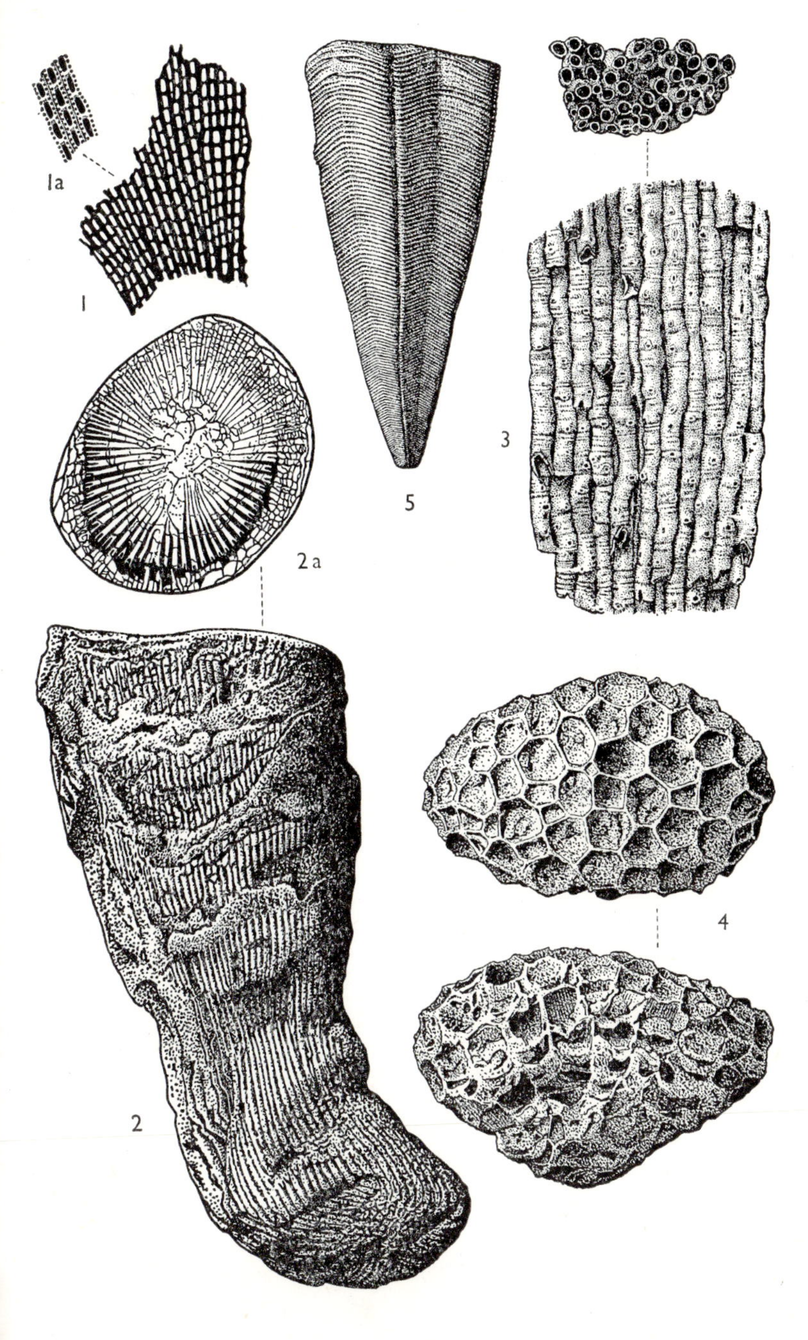

Plate 46

Carboniferous Brachiopods (Figs. 1–10) and Worm (Fig. 11)

1. **Overtonia fimbriata** (J. de C. Sowerby). (×1.) Viséan; Faulds Brow, Caldbeck, Cumbria. RANGE: Viséan–Namurian. [Syn., *Productus fimbriatus*.]

2–4. **Productus productus** (Martin). 2, 3 (×1.) 4 (×⅔.) Viséan; Beresford Hall, near Longnor, Staffordshire. RANGE: Genus, Viséan–Westphalian; Species, Viséan–Namurian.

5, 6. **Eomarginifera setosa** (Phillips). (×1¼.) Viséan. 5. Gorbeck, Settle, North Yorkshire. 6. Carluke, Lanarkshire. RANGE: Viséan.

7. **Orbiculoidea nitida** (Phillips). (×2.) Viséan; near Woodburn, Redesdale, Northumberland. RANGE: Genus, Ordovician–Permian Species, Viséan–Namurian.

8.* **Lingula mytilloides** J. Sowerby. (×1½.) Ammanian; Ystalyfera, near Swansea, West Glamorgan. RANGE: Genus, Ordovician–Recent; Species, Viséan–Westphalian.

9.* **Lingula squamiformis** Phillips. (×1½.) Viséan; Budle Bay, Bamburgh, Northumberland. RANGE: Genus, Ordovician–Recent; Species, Viséan–Namurian.

10. **Linoprotonia corrugatus** (M'Coy). (×¾.) (*a*), surface ornamentation (×1½.) Viséan; Settle, North Yorkshire. RANGE: Species, Viséan. [Syn., *Productus corrugatus*.]

11. **Spirorbis pusillus** (Martin). (×6.) Viséan Stage; Linhouse Water, Mid-Calder, Midlothian. RANGE: Genus, Silurian–Recent; Species, Carboniferous. [Syn., *Microconchus carbonarius* (Giebel).]

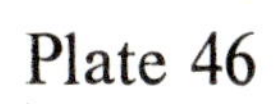

Plate 46

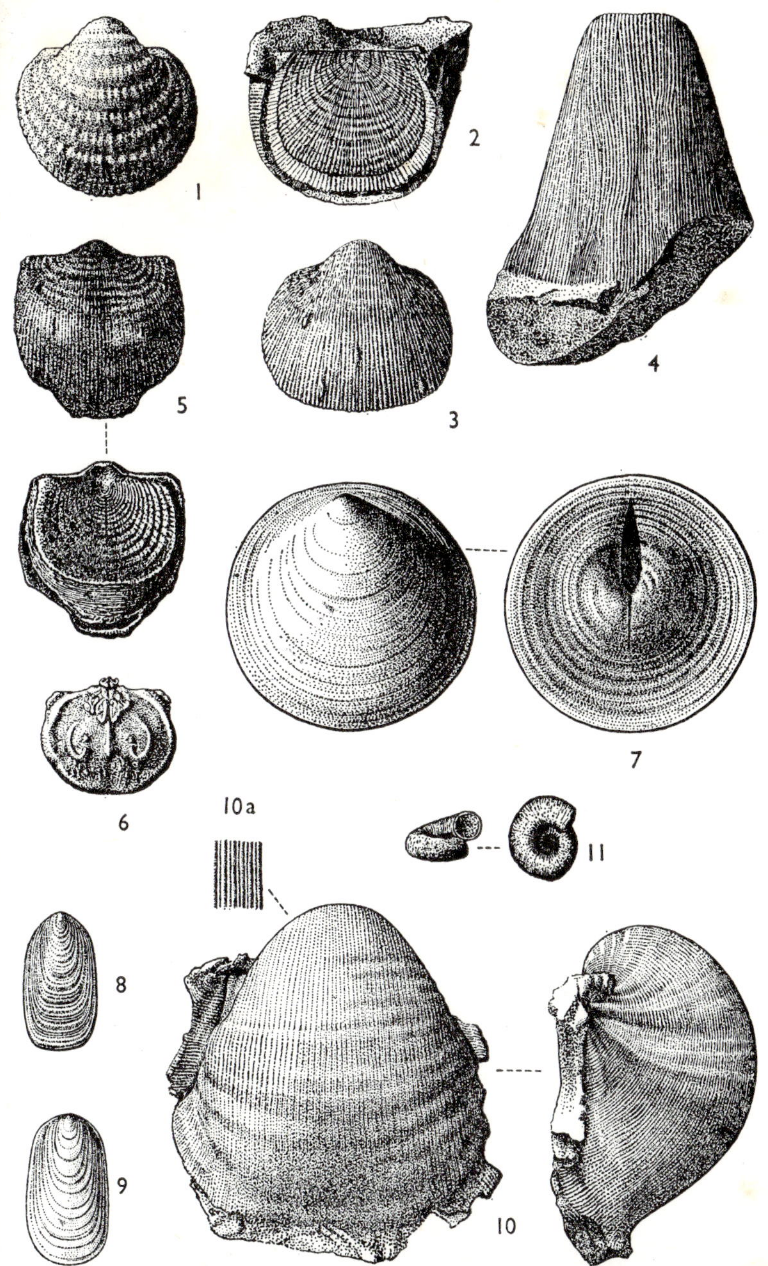

Plate 47

Carboniferous Brachiopods

1. **Rugosochonetes hardrensis** (Phillips). (×1½.) Viséan; Craignant, Chirk, near Wrexham, Clwyd. RANGE: Genus, Tournaisian–Namurian; Species, Viséan. [Syn., *Chonetes hardrensis*.]

2. **'Productus' craigmarkensis** (Muir-Wood). (×1½.) Westphalian; Smallthorne, Staffordshire. RANGE: Westphalian.

3. **Krotovia spinulosa** (J. Sowerby). (×1.) Lower Carboniferous; Faulds Brow, Caldbeck, Cumbria. RANGE: Genus, Lower Carboniferous–Permian; Species, Viséan–Namurian. [Syn., *Productus spinulosus*.]

4. **Antiquatonia hindi** (Muir-Wood). (×1.) Viséan Stage; Narrowdale, near Longnor, Staffordshire. RANGE: Genus, Carboniferous; Species, Viséan. [Syn., *Dictyoclostus hindi*, *Productus hindi*.]

5.* **Dictyoclostus semireticulatus** (Martin). (×¾.) Viséan; Bowland near Clitheroe, Lancashire. RANGE: Tournaisian–Namurian. [Syn., *Productus semireticulatus*.]

6.* **Gigantoproductus giganteus** (J. Sowerby). (×½.) Viséan; Llangollen, Clwyd. RANGE: Genus, Viséan–Namurian; Species, Viséan. [Syn., *Gigantella gigantea*, *Productus giganteus*.]

Plate 47

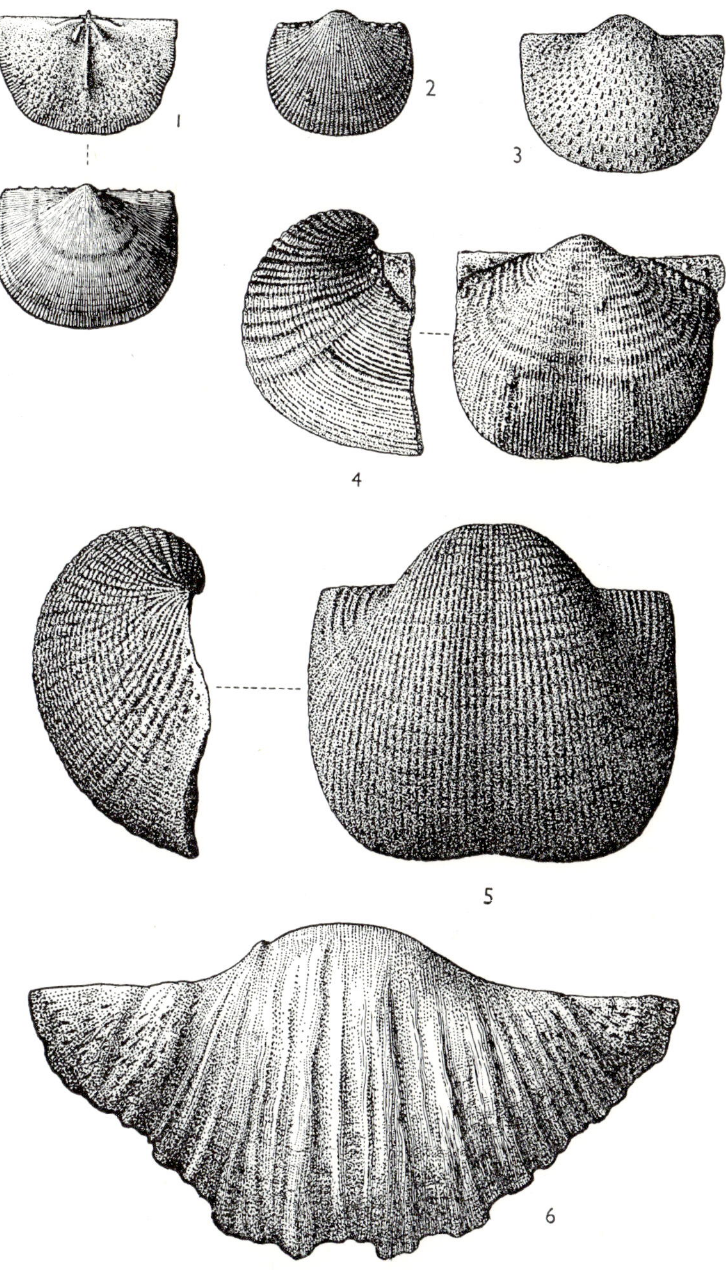

Plate 48

Carboniferous Brachiopods

1. **Leptagonia analoga** (Phillips). (×1.) 1. Viséan; Bowland, near Clitheroe, Lancashire. RANGE: Tournaisian–Namurian Stages. [Syn., *Leptaena analoga*, *Strophomena analoga*.]

2. **Leptagonia caledonica** Brand. 2 (×1.) Viséan; ironstone workings, Redesdale, Northumberland. RANGE: Genus, Middle Devonian–Namurian; Species, Viséan.

3, 4. **Pustula pustulosa** (Phillips). Carboniferous Limestone. 3 (×1.) Bowland, near Clitheroe, Lancashire. 4 (×1); 4*a* (×3.) Viséan; Narrowdale, near Longnor, Staffordshire. RANGE: Genus, Tournaisian–Viséan; Species, Viséan. [Syn., *Productus pustulosus*.]

5, 6. **Rhipidomella michelini** (Léveillé). (×1.) 5. Namurian; Congleton Edge, Cheshire. 6. Beith, Ayrshire. RANGE: Genus, Devonian–Permian; Species, Viséan–Namurian. [Syn., *Orthis michelini*.]

7, 8. **Martinothyris lineata** (J. Sowerby). 7 (×1.) Viséan; Whitewell, Settle, North Yorkshire. 8 (×10.) Ornamentation. RANGE: Genus, Lower Carboniferous–Permian; Species, Viséan–Namurian. [Syn., *Reticularia lineata*.]

Plate 48

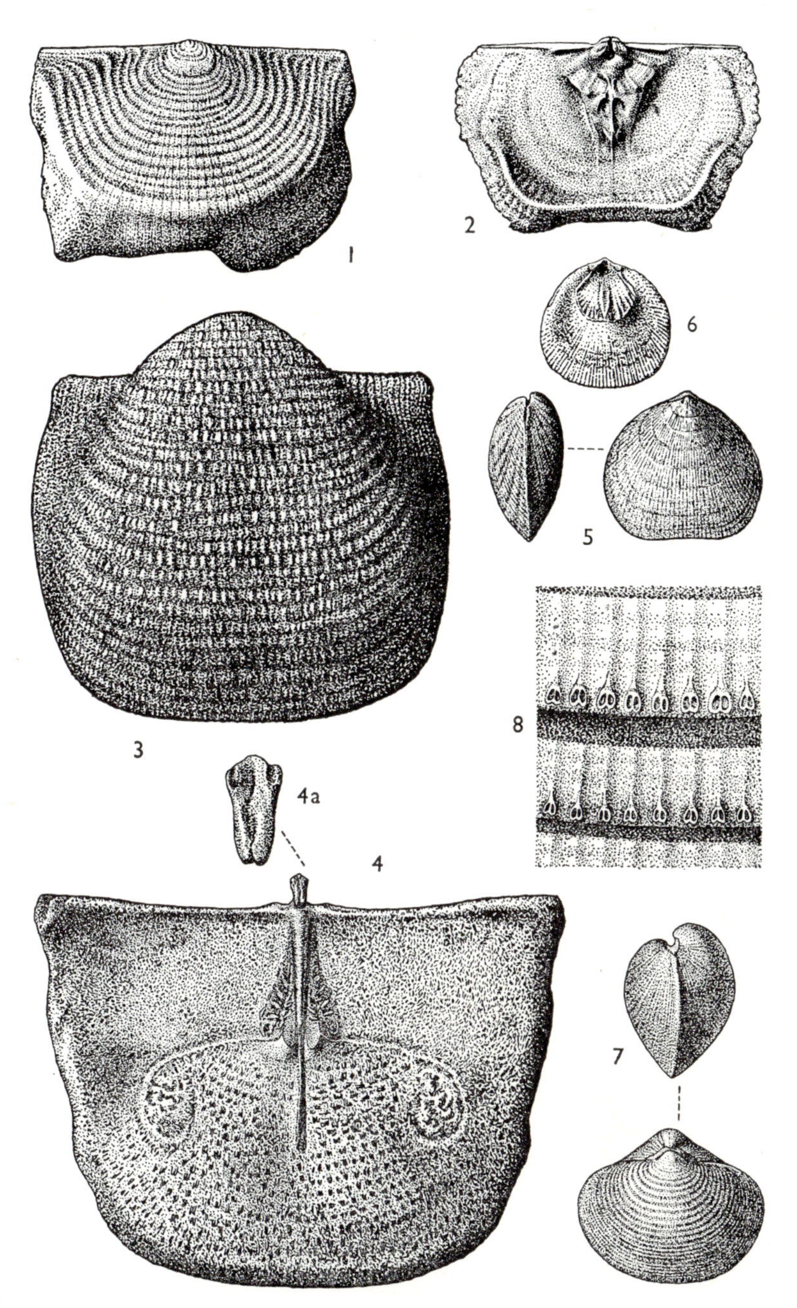

Plate 49

Carboniferous Brachiopods

1. **Schizophoria resupinata** (Martin). (×½.) Viséan; Narrowdale Hill, near Longnor, Staffordshire. RANGE: Genus, Devonian–Permian; Species, Viséan–Namurian. [Syn., *Orthis resupinata.*]

2.* **Schellwienella crenistria** (Phillips). (×1.) 2. Dorsal view. 2*a*. Ventral view of internal mould. Viséan; Elbolton, Yorkshire. RANGE: Genus, Tournaisian–Namurian; Species, Viséan–Namurian. [Syn., *Orthis crenistria.*]

3. **Brachythyris pinguis** (J. Sowerby). (×1.) Viséan; Millicent, County Kildare, Ireland. RANGE: Genus, Viséan–Westphalian; Species, Viséan. [Syn., *Spirifer pinguis.*]

4. **Daviesiella llangollensis** (Davidson). (×½.) Viséan; Llangollen, Clwyd. RANGE: Viséan. [Syn., *Productus llangollensis.*]

Plate 49

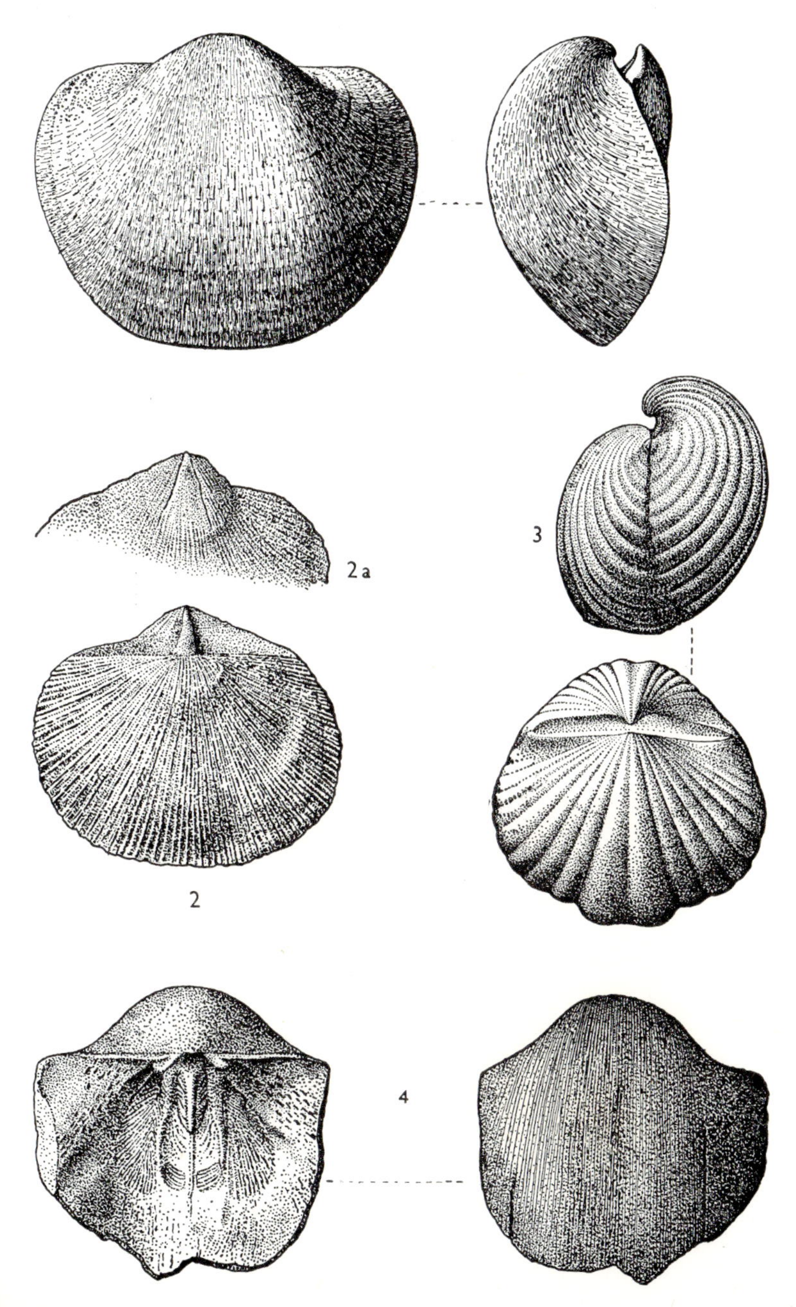

Plate 50

Carboniferous Brachiopods

1. **Composita ambigua** (J. Sowerby). (×1.) Lower Carboniferous; Settle, North Yorkshire. RANGE: Genus, Carboniferous, Viséan–Permian; Species, Viséan–Namurian. [Syn., *Athyris ambigua.*]

2. **Martinia glabra** (Martin). (×1.) Lower Carboniferous; Elden Hill, near Castleton, Derbyshire. RANGE: Viséan–Namurian. [Syn., *Spirifer glaber.*]

3.* **Spirifer striatus** (Martin). (×½.) Viséan; Derbyshire. RANGE: Genus, Carboniferous; Species, Tournaisian–Viséan.

Plate 50

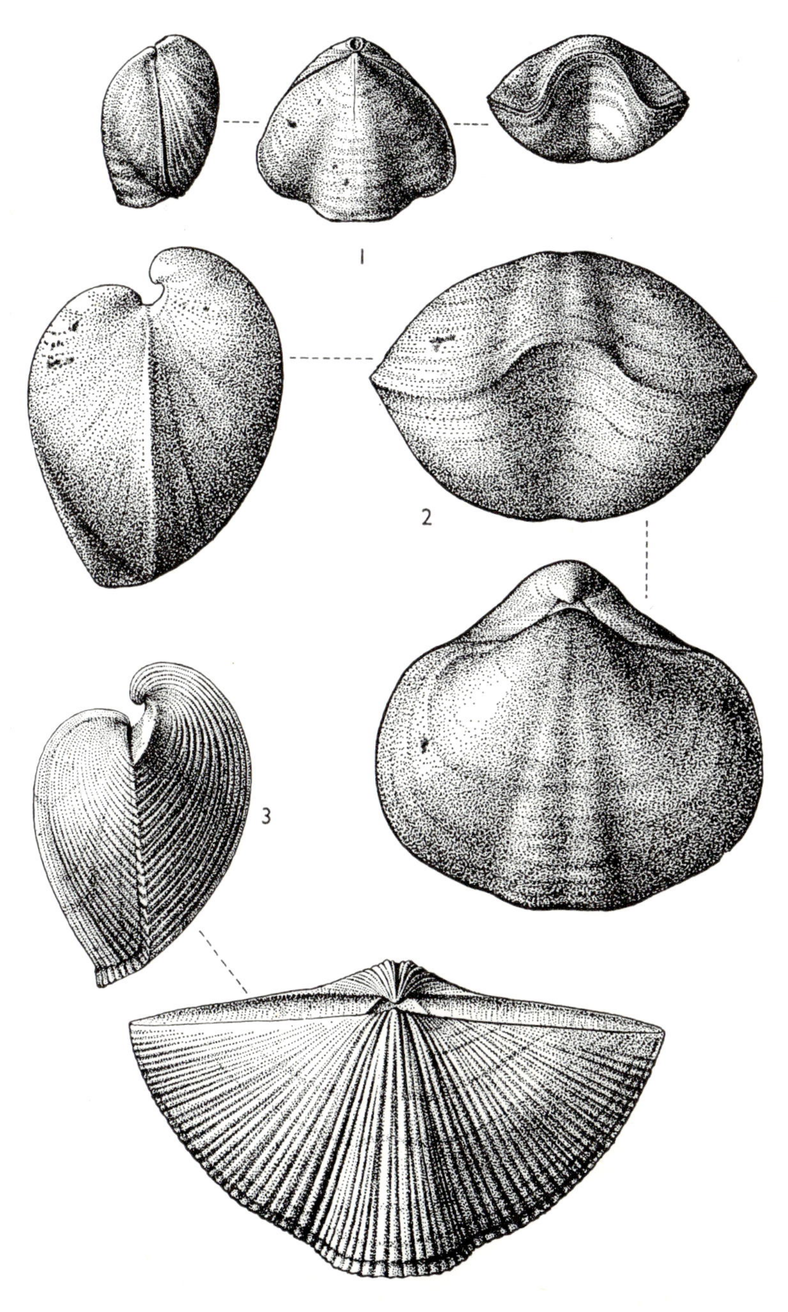

Plate 51

Carboniferous Brachiopods

1.* **Spirifer pennystonensis** George. (×1.) Upper Carboniferous; near Ironbridge, Salop. RANGE: Genus, Viséan–Westphalian; Species, Westphalian. [Syn., *Spirifer pennystonensis.*]

2. **Pleuropugnoides pleurodon** (Phillips). (×1½.) Viséan; Axton, near Prestatyn, Clwyd. RANGE: Viséan–Namurian. [Syn., *Pugnax pleurodon*, *Rhynchonella pleurodon.*]

3.* **Syringothyris cuspidata** (J. Sowerby). (×¾.) Lower Carboniferous; Rathkeale, County Limerick, Ireland. RANGE: Genus, Carboniferous; Species, Tournaisian–Viséan. [Syn., *Spirifer cuspidatus.*]

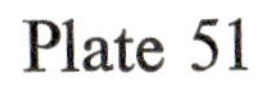
Plate 51

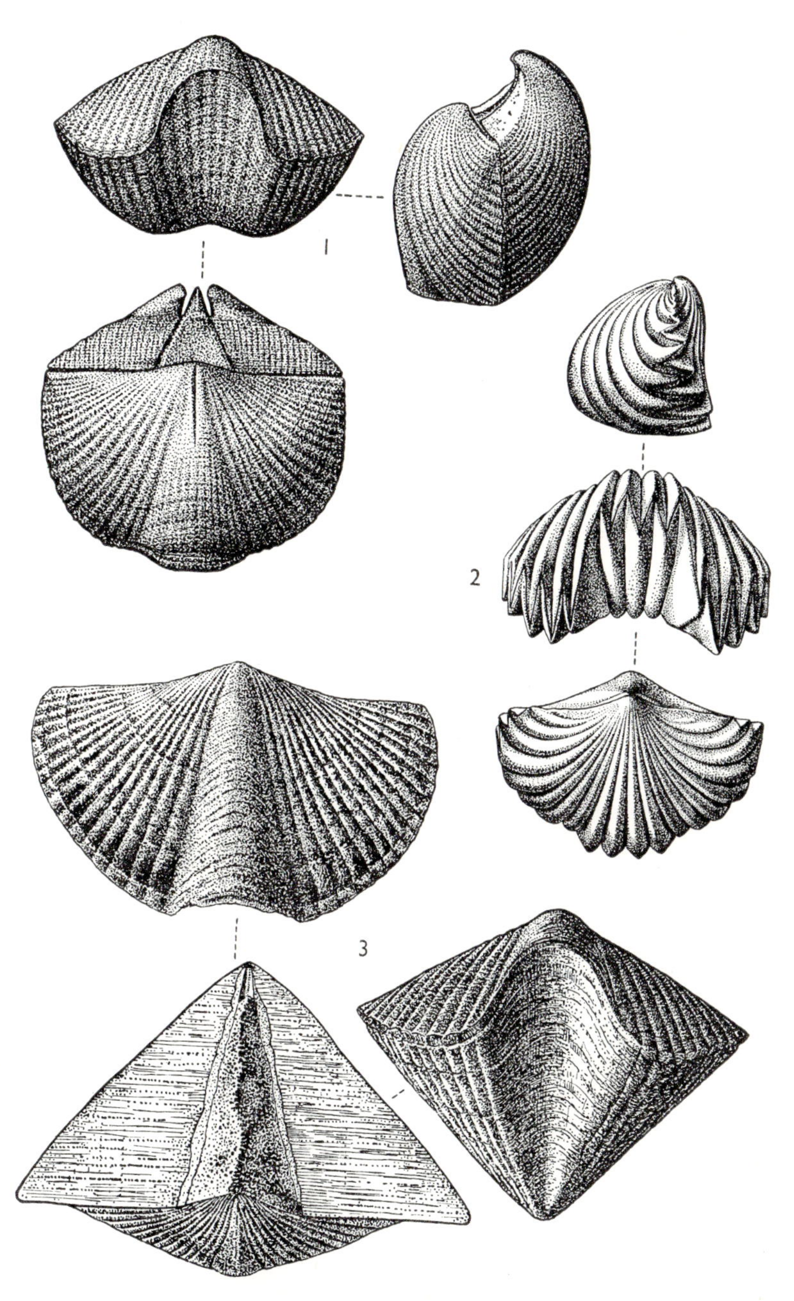

Plate 52

Carboniferous Brachiopods

1. **Punctospirifer scabricostus** North. (×2.) Viséan; Ash Fell, near Kirkby Stephen, Cumbria. RANGE: Genus, Carboniferous, Viséan Stage–Permian; Species, Viséan.

2. **Spirifer attenuatus** J. de C. Sowerby. (×1.) Lower Carboniferous; Kildare, Ireland. RANGE: Genus, Carboniferous; Species, Tournaisian.

3.* **Pugnax acuminatus** (J. Sowerby). (×1.) Viséan; Derbyshire. RANGE: Genus, Middle Devonian–Lower Carboniferous; Species, Viséan.

4. **Actinoconchus lamellosus** (Léveillé). (×1.) Showing marginal expansions. Viséan; Axton, near Prestatyn, Clwyd. RANGE: Tournaisian–Namurian. [Syn., *Athyris lamellosa*.]

5.* **Dielasma hastatum** (J. de C. Sowerby). (×1.) Carboniferous; Ireland. RANGE: Genus, Carboniferous, Viséan–Permian; Species, Viséan–Namurian. [Syn., *Terebratula hastata*.]

Plate 52

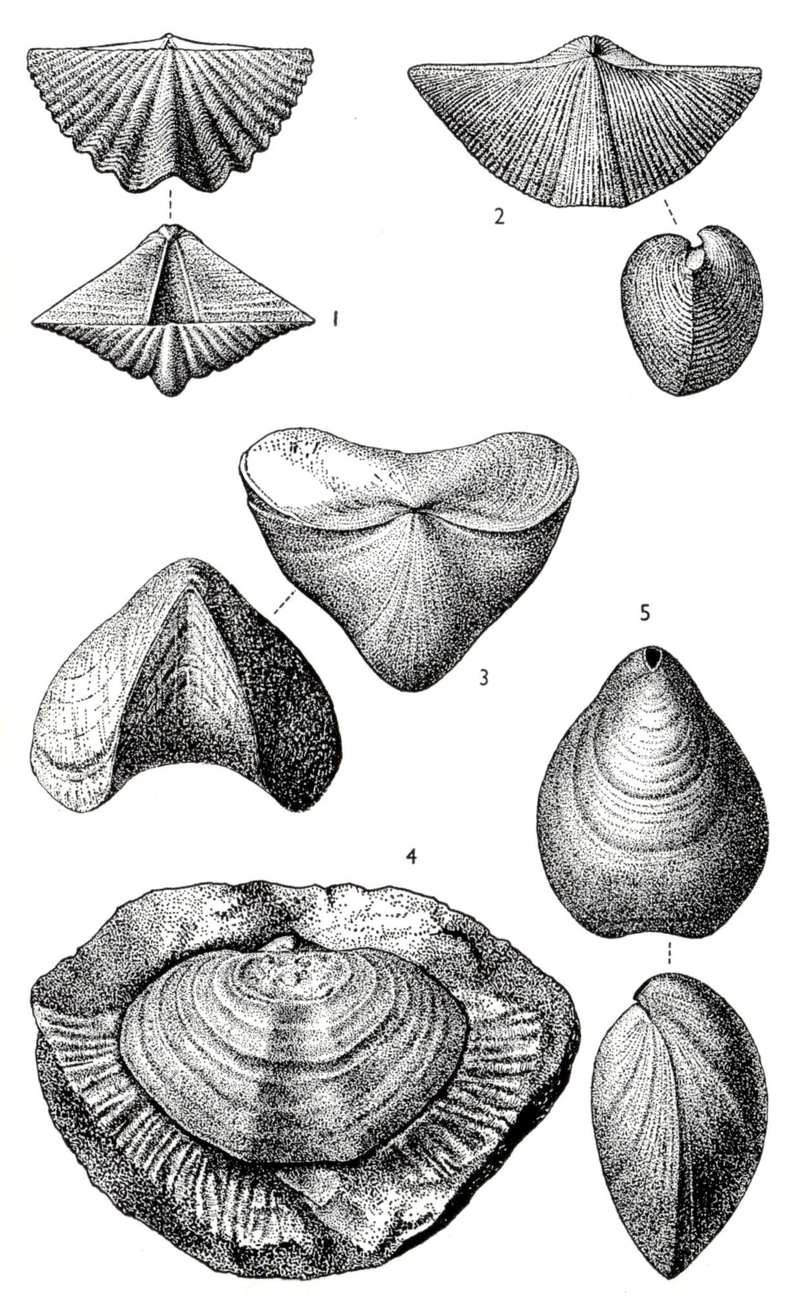

Plate 53

Carboniferous Bivalves (Figs. 1–4) and Rostroconchs (Figs. 5, 6)

1. **Polidevcia attenuata** (Fleming). (×1½.) Lower Carboniferous; Woodmill, near Dunfermline, Fifeshire. RANGE: Carboniferous. [Syn., *Nuculana attenuata*.]

2.* **Lithophaga lingualis** (Phillips). (×¾.) Viséan; Beith, Ayrshire. RANGE: Genus, Carboniferous–Recent; Species, Lower Carboniferous.

3. **Posidonia becheri** Bronn. (×1.) Lower Carboniferous; Budle, near Bamburgh, Northumberland. RANGE: Genus, Devonian–Carboniferous; Species, Lower Carboniferous. [Syn., *Posidonomya becheri*.]

4. **Posidoniella vetusta** (J. de C. Sowerby). (×1.) Viséan; near Castleton, Derbyshire. RANGE: Genus, Carboniferous; Species, Lower Carboniferous.

5, 6. **Conocardium hibernicum** J. Sowerby. Lower Carboniferous. 5 (×1.) Clitheroe, Lancashire. 6. Partly restored specimen (×½). Kildare, Ireland. **Conocardium** lacks the hinge and adductor muscle system typical of bivalves, the group in which until recently it was classified. RANGE: Genus, Ordovician–Permian; Species, Lower Carboniferous.

Plate 53

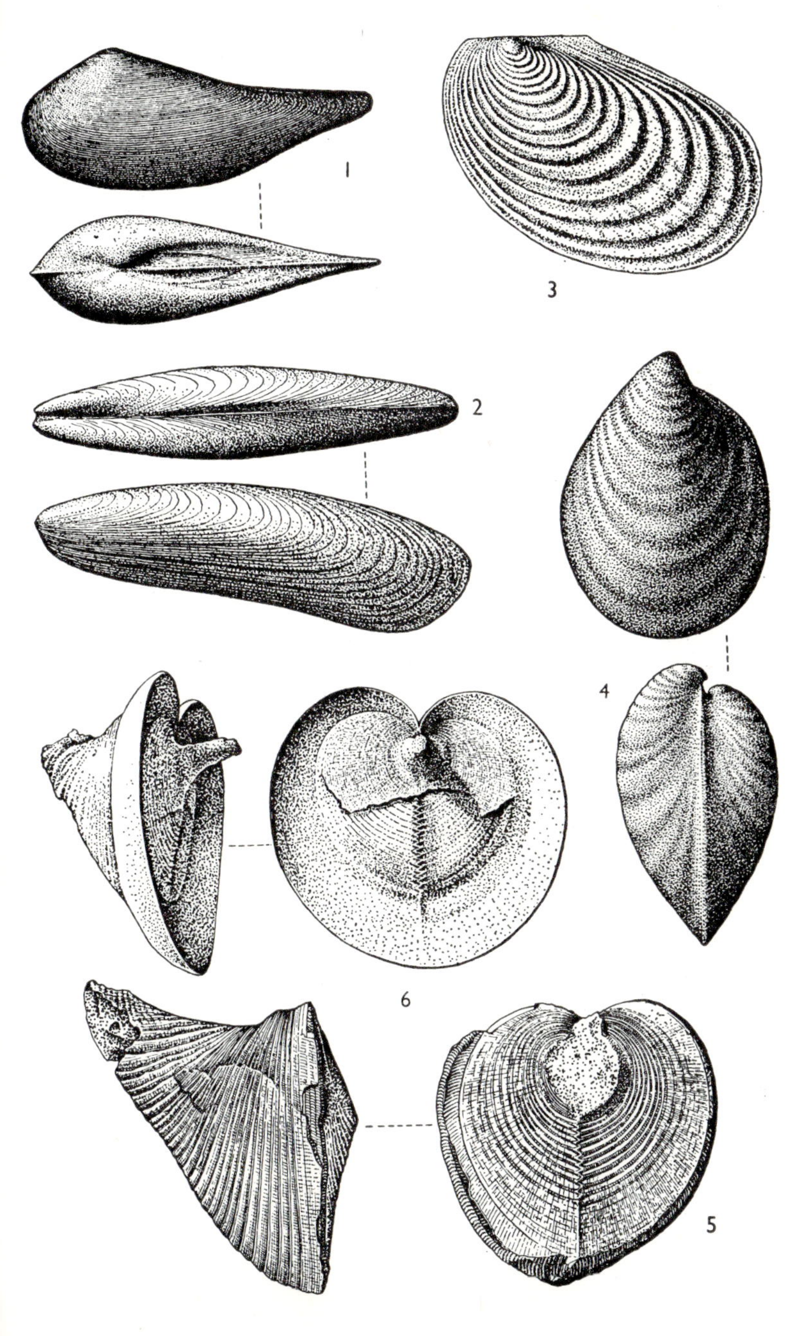

Plate 54

Carboniferous Marine Bivalves

1.* **Schizodus carbonarius** (J. de C. Sowerby). (×¾.) Upper Carboniferous, Coal Measures; Madeley, Salop. RANGE: Genus, Carboniferous–Permian; Species, Coal Measures.

2.* **Edmondia sulcata** (Phillips). (×¾.) Lower Carboniferous; Poolvash, near Castletown, Isle of Man. RANGE: Genus, Devonian–Permian; Species, Lower Carboniferous.

3.* **Wilkingia elliptica** (Phillips). (×1.) Upper Carboniferous, Coal Measures; Coalbrookdale, near Ironbridge, Salop. RANGE: Genus, Carboniferous–Permian; Species, Carboniferous. [Syn., *Allorisma sulcata* (Fleming).]

4. **Sanguinolites costellatus** M'Coy. (×1.) Lower Carboniferous, Viséan; Gurdy, near Beith, Ayrshire. RANGE: Genus, Devonian–Carboniferous; Species, Carboniferous.

5.* **Pterinopectinella granosa** (J. de C. Sowerby). (×1.) Lower Carboniferous; Kildare, Ireland. RANGE: Lower Carboniferous. [Syn., *Pterinopecten gransous* of authors.]

6.* **Dunbarella papyracea** (J. de C. Sowerby). (×1.) Upper Carboniferous, Coal Measures; Leeds. RANGE: Upper Carboniferous. [Syn., *Pterinopecten papyraceus*.]

7.* **Aviculopecten plicatus** (J. de C. Sowerby). (×1.) Lower Carboniferous; locality unknown. RANGE: Genus, Devonian ?–Carboniferous; Species, Lower Carboniferous.

Plate 54

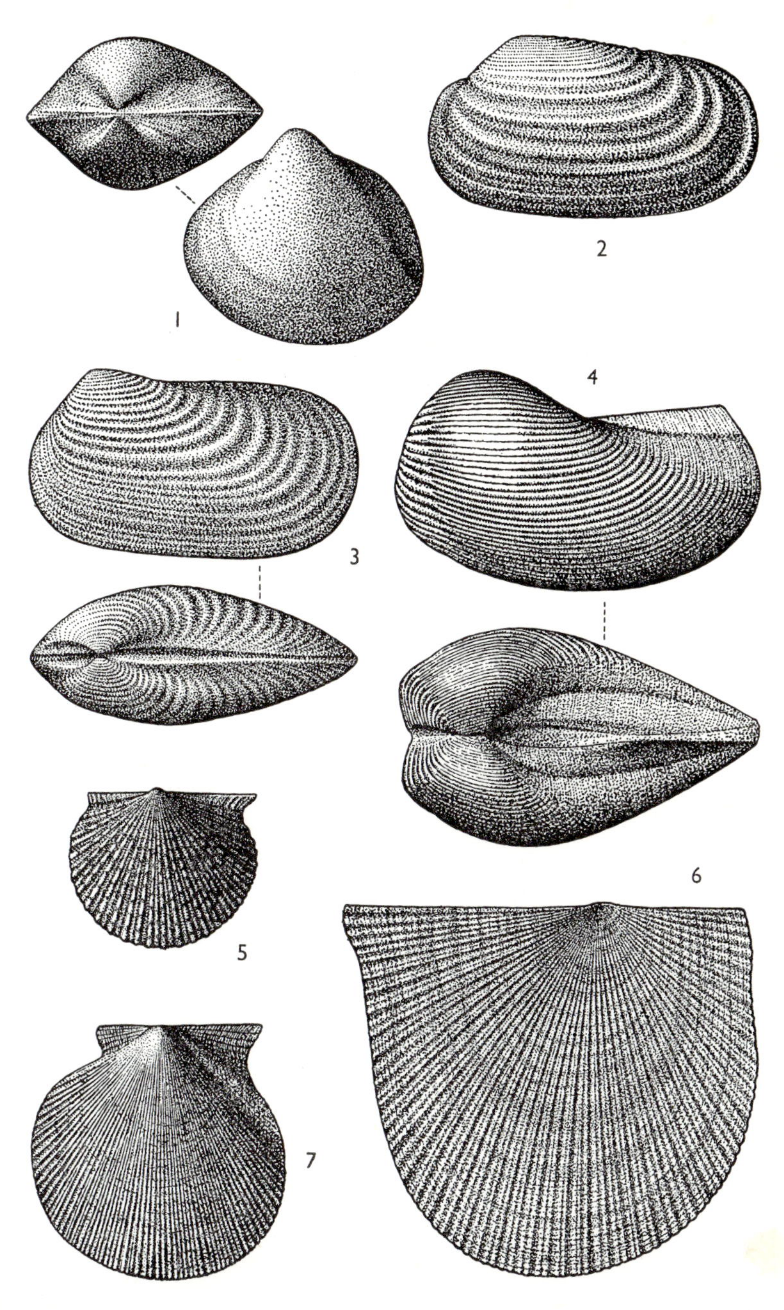

Plate 55

Carboniferous Non-marine Bivalves

1.* **Naiadites modiolaris** J. de C. Sowerby. (×1.) Coal Measures, Ammanian Stage; Adderley Green, Longton, Stoke-on-Trent, Staffordshire. RANGE: Genus, Carboniferous, Tournaisian–Ammanian Stages; Species, Ammanian Stage, *modiolaris*–Lower *similis–pulchra* Zones.

2. **Anthracosphaerium exiguum** (Davies & Trueman). (×1.) Coal Measures, Ammanian Stage; Ystalyfera, near Swansea, West Glamorgan. RANGE: Genus, Carboniferous, Ammanian–Stephanian Stages; Species, Ammanian Stage, *modiolaris*–Lower *similis–pulchra* Zones.

3. **Anthracosia atra** (Trueman). (×1.) Coal Measures, Ammanian Stage; Ystalyfera, near Swansea, West Glamorgan. RANGE: Genus, Carboniferous, Ammanian Stage; Species, Lower *similis–pulchra* Zone.

4. **Anthracosia planitumida** (Trueman). (×1½.) Coal Measures, Ammanian Stage; Ystalyfera, near Swansea, West Glamorgan. RANGE: Genus, Carboniferous, Ammanian Stage; Species, Lower *similis–pulchra* Zone.

5.* **Carbonicola communis** Davies & Trueman. (×1.) Coal Measures, Ammanian Stage; Black Mountain Colliery, Twrch Valley, Dyfed. RANGE: Genus, Carboniferous, Tournaisian–Ammanian; Species, Ammanian Stage, *communis* Zone.

6.* **Carbonicola pseudorobusta** Trueman. (×½.) Coal Measures, Ammanian Stage; Halifax, West Yorkshire. RANGE: Genus, Carboniferous, Tournaisian–Ammanian; Species, Ammanian Stage, *communis* Zone.

7.* **Anthraconaia adamsi** (Salter). (×¾.) Coal Measures, Ammanian Stage; near Fenton, Stoke-on-Trent, Staffordshire. RANGE: Genus, Carboniferous, Ammanian–Stephanian Stages; Species, Ammanian Stage, Upper *similis–pulchra* Zone.

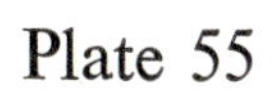

Plate 55

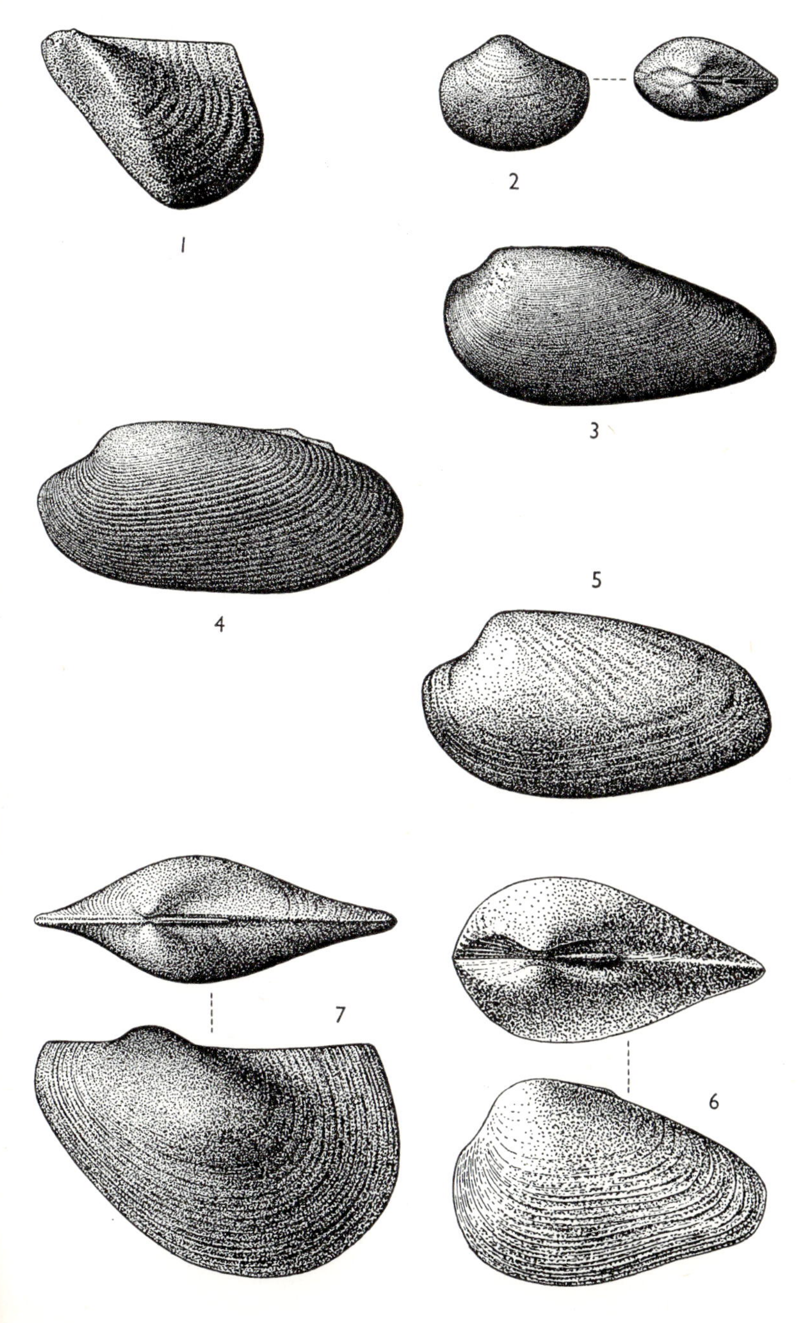

Plate 56

Carboniferous Monoplacophoran (Fig. 1) and Gastropods (Figs. 2–9)

1.* **Euphemites urii** (Fleming). (×1½.) Lower Carboniferous; Gare, Carluke, Lanarkshire. RANGE: Genus, Carboniferous; Species, Lower Carboniferous. [Syn., *Euphemus urii.*]

2.* **Euconospira conica** (Phillips). (×1.) 2*a*, Surface ornamentation (×2.) Lower Carboniferous; Bowland, near Clitheroe, Lancashire. RANGE: Lower Carboniferous. [Syn., *Mourlonia conica.*]

3.* **Straparollus dionysii** Montfort. (×¾.) Lower Carboniferous, Viséan; Wedber Knoll, near Malham, North Yorkshire. RANGE: Genus, Devonian–Carboniferous; Species, Lower Carboniferous.

4. **Glabrocingulum atomarium** (Phillips). (×2¼.) 4*a*, Surface ornamentation (×5.) Viséan; Woodmill, near Dunfermline, Fifeshire. RANGE: Genus, Viséan–Namurian; Species, Viséan Stage.

5. **Platyceras vetustum** (J. de C. Sowerby). (×1.) Lower Carboniferous; Ireland. RANGE: Genus, Silurian–Carboniferous; Species, Lower Carboniferous.

6.* **Soleniscus acutus** (J. de C. Sowerby). (×¾.) Lower Carboniferous; Kildare, Ireland. RANGE: Genus, Lower Carboniferous–Permian; Species, Lower Carboniferous. [Syn., *Macrochilina acuta.*]

7. **Mourlonia carinata** (J. Sowerby). (×½.) Lower Carboniferous; Bowland, near Clitheroe, Lancashire. RANGE: Genus, Devonian–Permian; Species, Lower Carboniferous.

8.* **Straparollus pentangulatus** (J. Sowerby). (×1.) Lower Carboniferous; Bowland, near Clitheroe, Lancashire. RANGE: Genus, Devonian–Carboniferous; Species, Lower Carboniferous. [Syn., *Euomphalus pentangulatus.*]

9. **Glabrocingulum armstrongi** Thomas. (×2.) *a*, Surface ornamentation (×5.) Lower Carboniferous; Wilkieston, Kirknewton, Midlothian. RANGE: Viséan–Namurian.

Plate 56

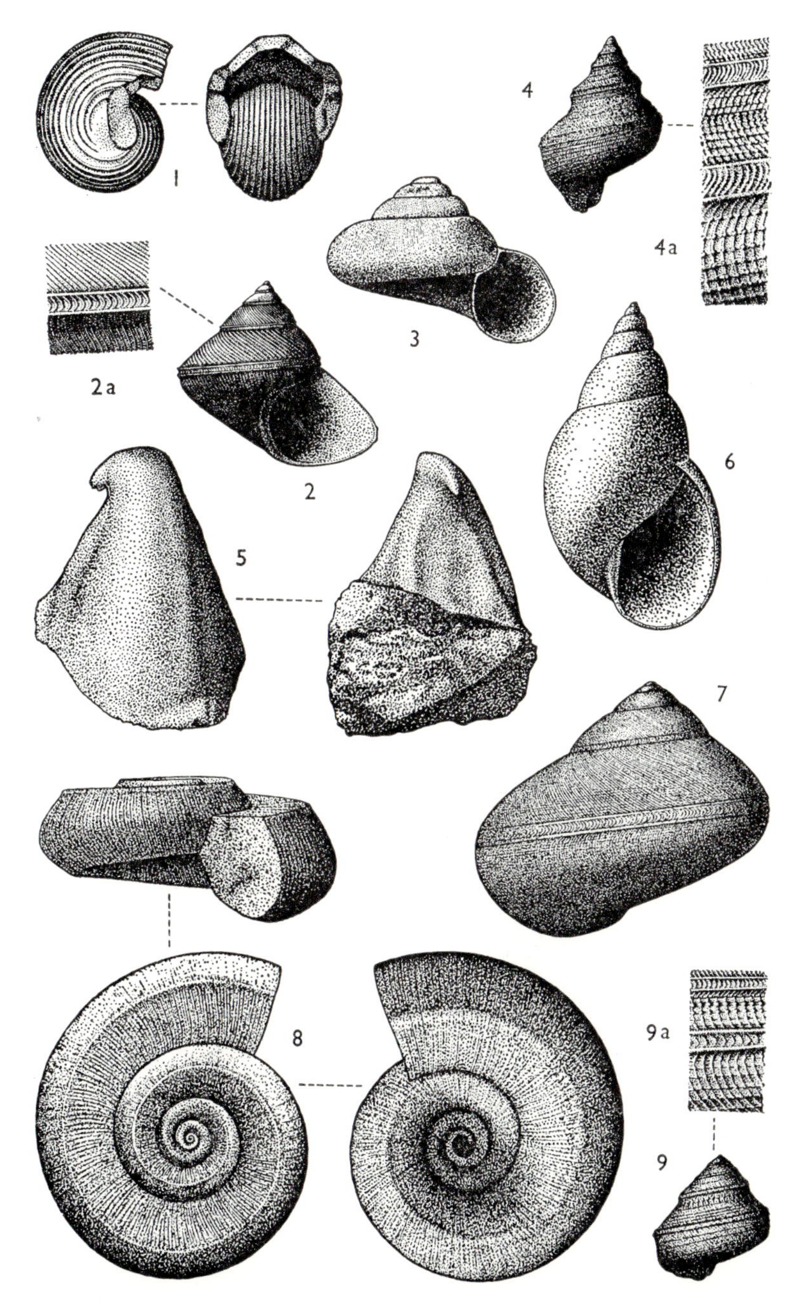

Plate 57

Carboniferous Gastropods (Figs. 1, 2) and Goniatites (Figs. 3, 4)

1.* **Naticopsis elliptica** (Phillips). ($\times\frac{3}{4}$.) Lower Carboniferous; Kildare, Ireland, RANGE: Genus, Devonian–Permian (? Trias); Species, Lower Carboniferous.

2.* **Palaeostylus rugiferus** (Phillips). ($\times$1.) Lower Carboniferous; Craigenglen, Campsie, Stirlingshire. RANGE: Carboniferous. [Syn., *Zygopleura rugifera*.]

3. **Reticuloceras reticulatum** (Phillips). ($\times 1\frac{1}{2}$.) Namurian; Hebden Bridge, North Yorkshire. RANGE: Genus, Namurian, Zone R; Species, Zone R_1.

4. **Reticuloceras bilingue** (Salter). ($\times$1.) 4*a*, Ornamentation ($\times 2\frac{1}{2}$.) Namurian; Hebden Bridge, North Yorkshire. RANGE: Genus, Namurian, Zone R; Species, Zone R_2. [Syn., *Reticuloceras reticulatum* mut. β.]

Plate 57

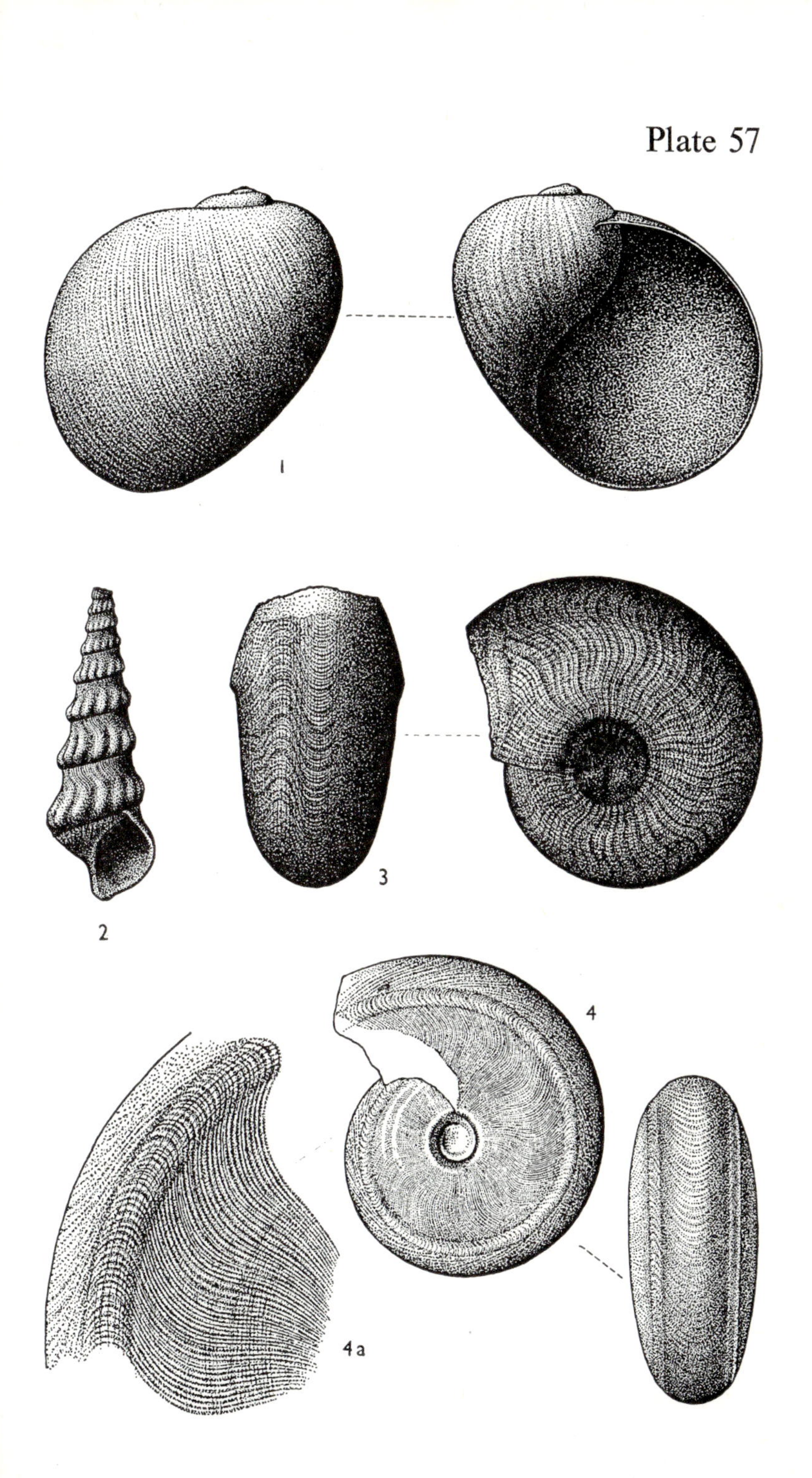

Plate 58

Carboniferous Goniatites

1.* **Homoceras diadema** (Beyrich). (×1½.) Namurian; West Yorkshire. RANGE: Namurian, Zone H.

2, 3.* **Muensteroceras truncatum** (Phillips). (×1.) 3, Septal suture. Viséan; Bowland, near Clitheroe, Lancashire. RANGE: Viséan, Zone P. [Syn., *Beyrichoceratoides truncatum.*]

4, 5. **Beyrichoceras obtusum** (Phillips). (×1.) 5, Septal suture. Viséan; Bowland, near Clitheroe, Lancashire. RANGE: Viséan; Genus, Zones B_1–P_1; Species, Zone P_1.

6. **Gastrioceras carbonarium** (Buch). (×1.) Upper Carboniferous; Churnet Valley, near Leek, Staffordshire. RANGE: Namurian–Ammanian. Genus, Zone G; Species, Zone G_2.

Plate 58

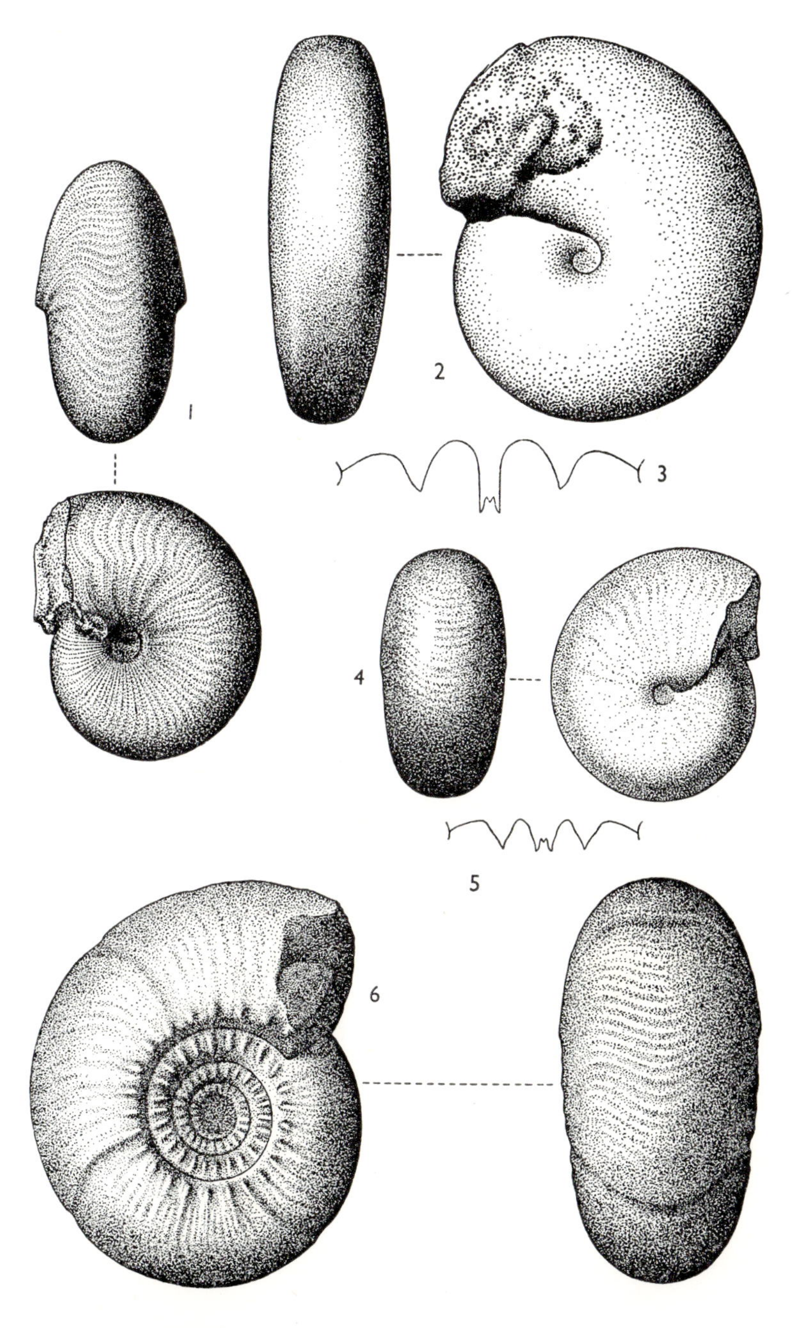

Plate 59

Carboniferous Goniatites (Figs. 1, 2), Blastoids (Figs. 3–7) and Echinoids (Figs. 8–10)

1.* **Goniatites crenistria** Phillips. (×1.) *a*, septal suture. Lower Carboniferous; Dinckley Ferry, River Ribble, near Blackburn, Lancashire. RANGE: Genus, Viséan; Species, Zone P.

2.* **Neoglyphioceras spirale** (Phillips). (×1.) Viséan; Waldon Barton, near Chudleigh, Devon. RANGE: Viséan. [Syn., *Goniatites spiralis*.]

3, 4.* **Codaster acutus** M'Coy. (×1½.) 3, upper surface. 4, side view. Lower Carboniferous; Settle, North Yorkshire. RANGE: Lower Carboniferous. [Syn., *Codastertrilobatus* M'Coy, *Codonaster trilobatus*.]

5, 6. **Orbitremites ellipticus** (G. B. Sowerby). (×1½.) Viséan; Lancashire. RANGE: Genus, Carboniferous; Species, Viséan. [Syn., *Granatocrinus ellipticus*.]

7. **Orophocrinus verus** (Cumberland). (×1½.) Viséan; Whitewell, Bowland, near Clitheroe, Lancashire. RANGE: Genus, Lower Carboniferous; Species, Viséan.

8. **Archaeocidaris** sp. Radiole (×1.) Lower Carboniferous; Chrome Hill, near Buxton, Derbyshire. RANGE: Genus, Carboniferous.

9. **Archaeocidaris urii** (Fleming). Plate (×3.) Lower Carboniferous; Roscobie, near Dunfermline, Fifeshire. RANGE: Species, Tournaisian ?–Viséan. [Syn., *Cidaris benburbensis* Portlock.]

10. **Lovenechinus lacazei** (Julien). (×¾.) Viséan; near Kirkby Stephen, Cumbria. RANGE: Genus, Tournaisian–Viséan; Species, Viséan. [Syn., *Palaeechinus sphaericus* Koninck *non* M'Coy.]

Plate 59

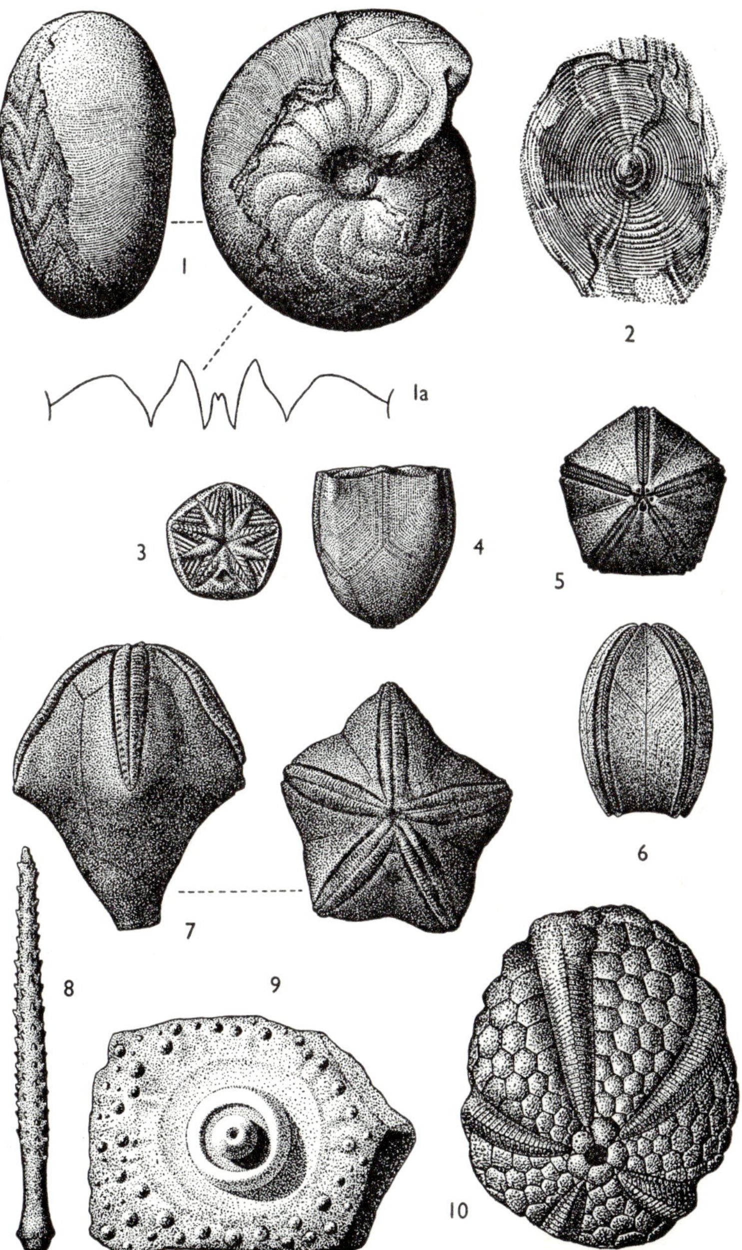

Plate 60

Carboniferous Echinoid (Fig. 1) and Crinoids (Figs. 2–4)

1.* **Melonechinus etheridgei** (Keeping). Fragments of test (×1.) Lower Carboniferous; Thor's Cave, Manifold Valley, near Wetton, Staffordshire. RANGE: Lower Carboniferous.

2.* **Woodocrinus macrodactylus** Koninck. (×1.) Namurian; Richmond, North Yorkshire. RANGE: Genus, Carboniferous, Viséan–Namurian; Species, Namurian.

3. **Amphoracrinus gigas** Wright. (×¾.) Tournaisian; Balnaleck, near Florence Court, Enniskillen, Co. Fermanagh, Northern Ireland. RANGE: Genus, Lower Carboniferous; Species, Tournaisian.

4.* **Platycrinites gigas** Phillips. (×1¼.) Viséan; Bowland, near Clitheroe, Lancashire. RANGE: Genus, Devonian–Carboniferous; Species, Viséan.

Plate 60

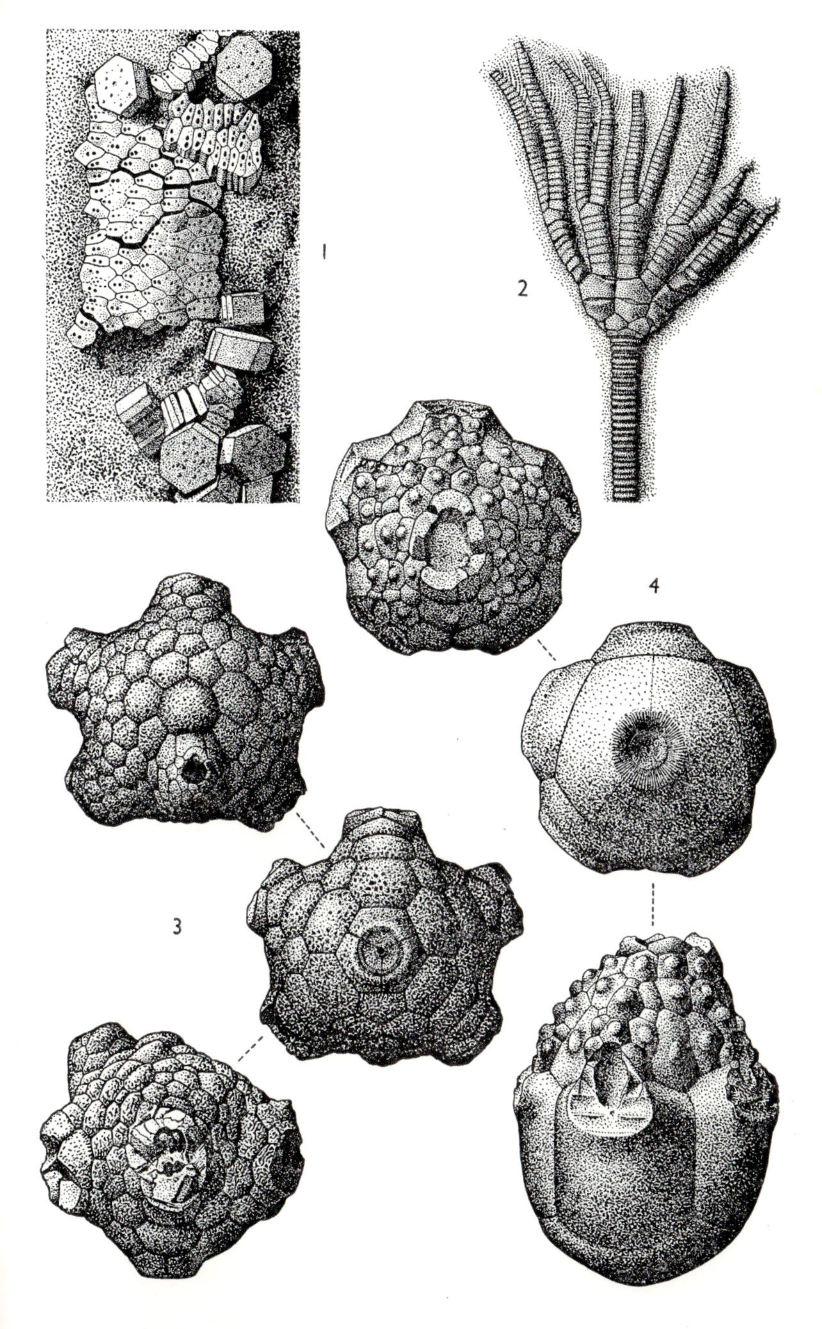

Plate 61

Carboniferous Crinoids

1.* **Actinocrinites triacontadactylus** Miller. (×1.) *a*, tegminal; *b*, basal; *c*, side view. Lower Carboniferous; locality unknown. RANGE: Genus, Carboniferous; Species, Viséan.

2.* **Gilbertsocrinus konincki** Grenfell. (×1.) *a*, tegminal, *b*, basal, *c*, side view. Lower Carboniferous; Yorkshire. RANGE: Genus, Middle Devonian–Lower Carboniferous; Species, Viséan. [Syn., *Gilbertsocrinus calcaratus* Bather *non* Phillips.]

Plate 61

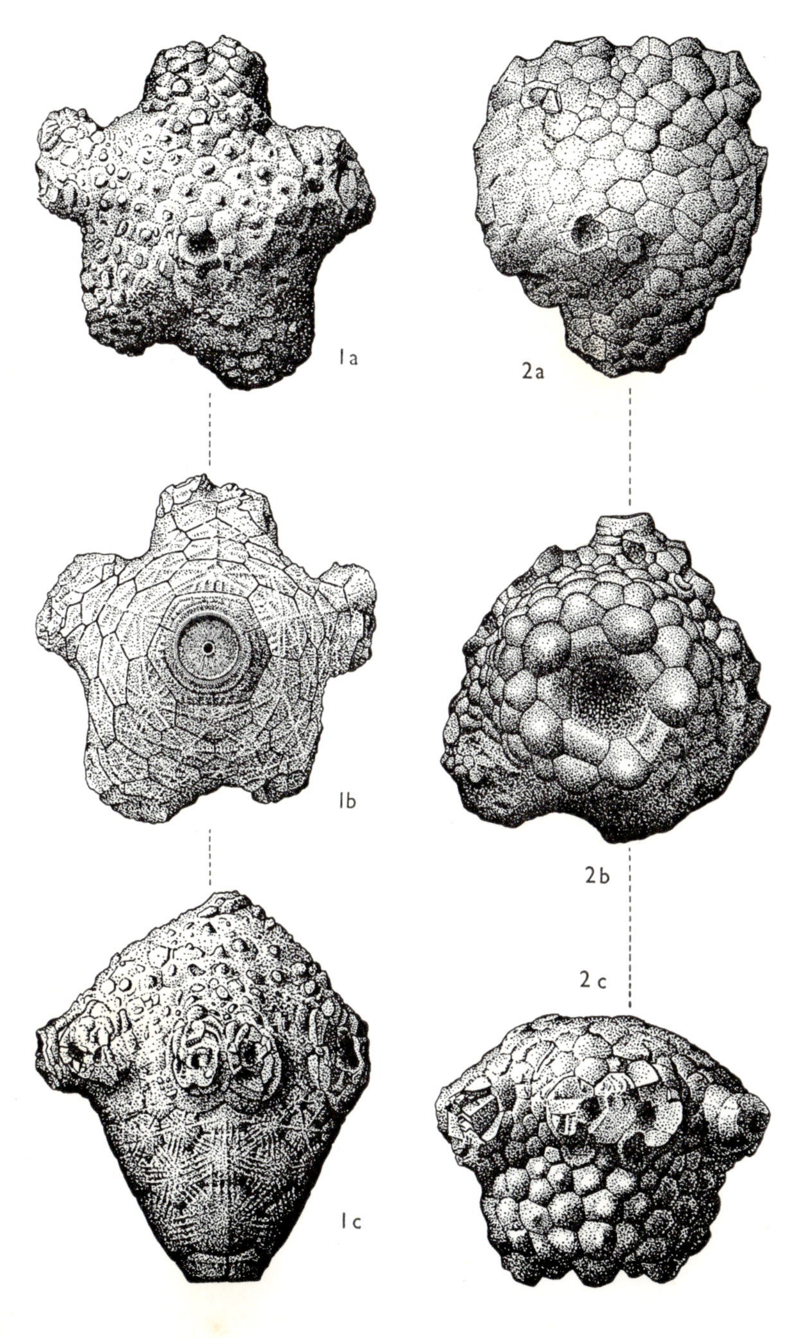

Plate 62

Carboniferous Trilobites (Figs. 1–9) and Crustacean (Fig. 10)

1.* **Eocyphinium seminiferum** (Phillips). (×1.) Lower Carboniferous, Viséan; Matlock, Derbyshire. RANGE: Lower Carboniferous. [Syn., *Griffithides seminiferus*.]

2. **Cummingella jonesi** (Portlock). (×1.) Lower Carboniferous; Bowland, near Clitheroe, Lancashire. RANGE: Lower Carboniferous. [Syn., *Phillipsia derbiensis* of authors.]

3, 4.* **Brachymetopus ouralicus** (Verneuil). (×2.) Lower Carboniferous. 3, Viséan; Peakshill Farm, Mam Tor, near Castleton, Derbyshire. 4, Derbyshire. RANGE: Lower Carboniferous.

5. **Phillipsia gemmulifera** (Phillips). (×2.) Lower Carboniferous; Bowland, near Clitheroe, Lancashire. RANGE: Lower Carboniferous. [Syn., *Asaphus gemmuliferus*.]

6, 7.* **Spatulina spatulata** (Woodward). (×1½.) Lower Carboniferous, Viséan; Coddon Hill, near Barnstaple, Devonshire. RANGE: Viséan. [Syn., *Phillibole* (*Cystispina*) *spatulata*, *Phillipsia spatulata*.]

8, 9. **Cummingella jonesi** (Portlock). (×4.) Lower Carboniferous; Bowland, near Clitheroe, Lancashire. RANGE: Lower Carboniferous. [Syn., *Phillipsia laticaudata*.]

10.* **Perimecturus parki** (Peach). (×1.) Lower Carboniferous, Viséan, Calciferous Sandstone Series; near Langholm, Eskdale, Dumfriesshire. RANGE: Genus, Carboniferous; Species, Lower Carboniferous.

Plate 62

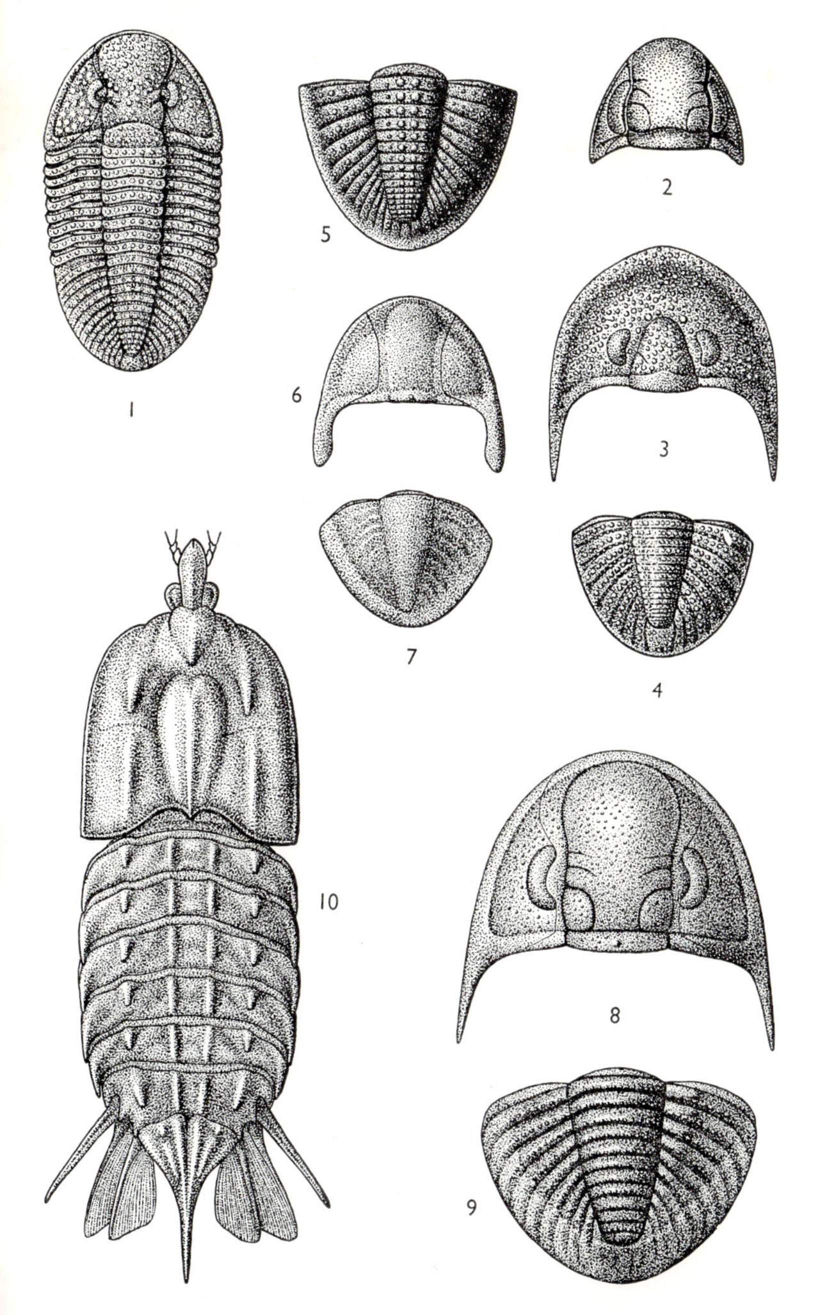

Plate 63

Carboniferous Arthropods (Figs. 1–6), Conodonts (Figs. 7, 8) and Fish (Fig. 9)

1. **Entomoconchus scouleri** M'Coy. (×1.) Lower Carboniferous, Viséan; Clitheroe, Lancashire. RANGE: Lower Carboniferous.

2. **Richteria biconcentrica** (Jones). (×4.) Lower Carboniferous; Little Island, near Cork, Ireland. RANGE: Genus, Devonian–Lower Carboniferous; Species, Lower Carboniferous. [Syn., *Entomis biconcentrica*.]

3. **Ectodemites bipartitus** (Vine). (×25.) Lower Carboniferous, Namurian; Orchard, near Glasgow. RANGE: Genus, Carboniferous–Permian; Species, Lower Carboniferous. [Syn., *Kirkbya bipartita*.]

4.* **Euproops rotundatus** (Prestwich). (×¾.) Upper Carboniferous, Coal Measures; Coseley, Staffordshire. RANGE: Upper Carboniferous. [Syn., *Prestwichia rotundata*, *Prestwichianella rotundata*.]

5.* **Acantherpestes ferox** (Salter). (×1.) Upper Carboniferous, Coal Measures; Coseley, Staffordshire. This is part of the body of a giant millipede. RANGE: Upper Carboniferous.

6. **Eophrynus prestvici** (Buckland). (×2.) Upper Carboniferous, Coal Measures; Dudley, West Midlands. RANGE: Upper Carboniferous.

7. **Idiognathoides corrugata** (Harris & Hollingworth). (×35.) Upper Carboniferous, Namurian Stage; Oakamoor, near Cheadle, Staffordshire. RANGE: Namurian.

8. **Gnathodus bilineatus** (Roundy). (×18.) Upper Carboniferous, Namurian Stage; Waterhouses, near Leek, Staffordshire. RANGE: Viséan–Namurian.

9.* **Cladodus mirabilis** Agassiz. Tooth (×1.) Lower Carboniferous; Armagh, N. Ireland. RANGE: Genus, Carboniferous; Species, Lower Carboniferous.

Plate 63

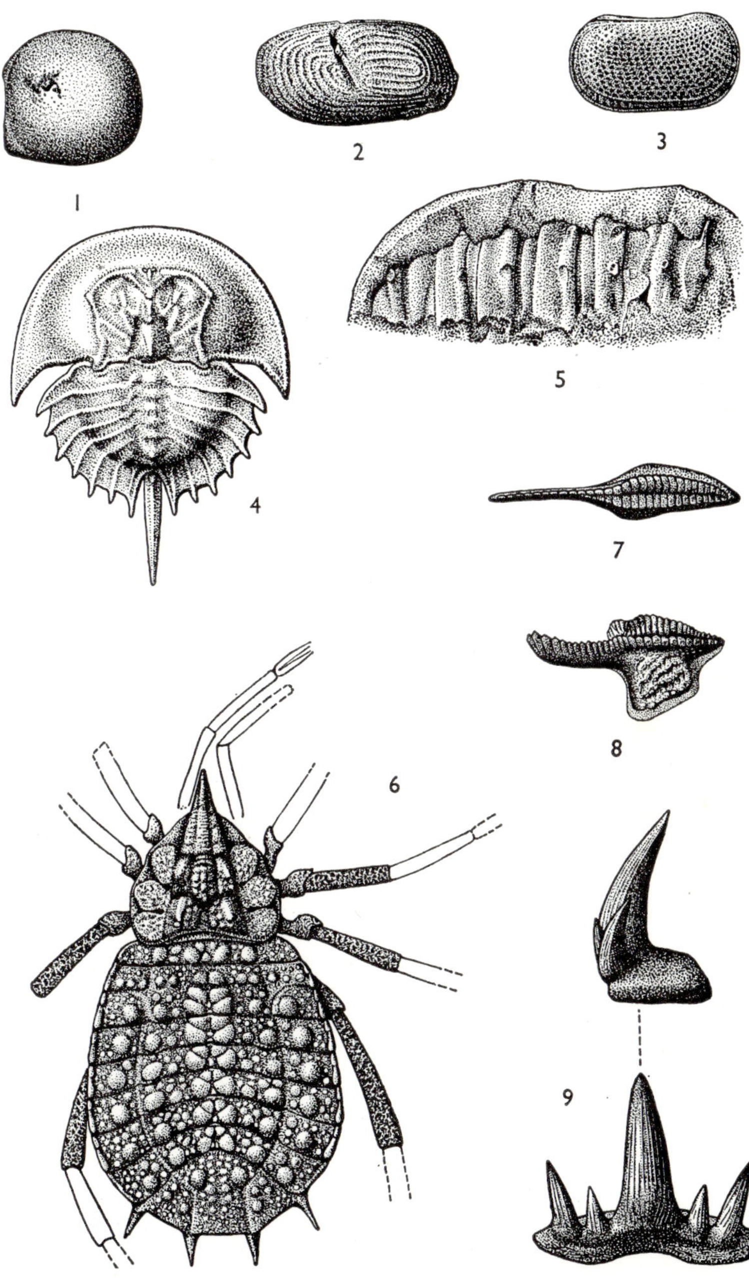

Plate 64

Carboniferous Fishes

1.* **Psephodus magnus** (Portlock). Tooth-plate (× 1.) Lower Carboniferous; Armagh, N. Ireland. RANGE: Genus, Carboniferous; Species, Lower Carboniferous.

2.* **Psammodus regosus** Agassiz. Tooth-plate (× $\frac{3}{4}$.) *2a,* surface (× 10.) Lower Carboniferous; Armagh, N. Ireland. RANGE: Lower Carboniferous.

3.* **Helodus turgidus** (Agassiz). Tooth-plate (× 1.) Lower Carboniferous; Bristol. RANGE: Genus, Carboniferous; Species, Lower Carboniferous.

4.* **Orodus ramosus** Agasisz. Tooth (× $\frac{3}{4}$.) Lower Carboniferous; Oreton, near Ludlow, Salop. RANGE: Genus and Species, Lower Carboniferous.

5.* **Gyracanthus formosus** Agassiz. Fin-spine (× $\frac{1}{2}$.) Coal Measures; Dalkeith, Midlothian. RANGE: Genus and Species, Carboniferous.

6.* **Xenacanthus laevissimus** Agassiz). Head-spine (× $\frac{1}{2}$.) *a,* margin (× 1$\frac{1}{2}$.) Upper Carboniferous; Dalkeith, Midlothian. RANGE: Genus, Carboniferous; Species, Upper Carboniferous. [Syn., *Pleuracanthus laevissimus.*]

Plate 64

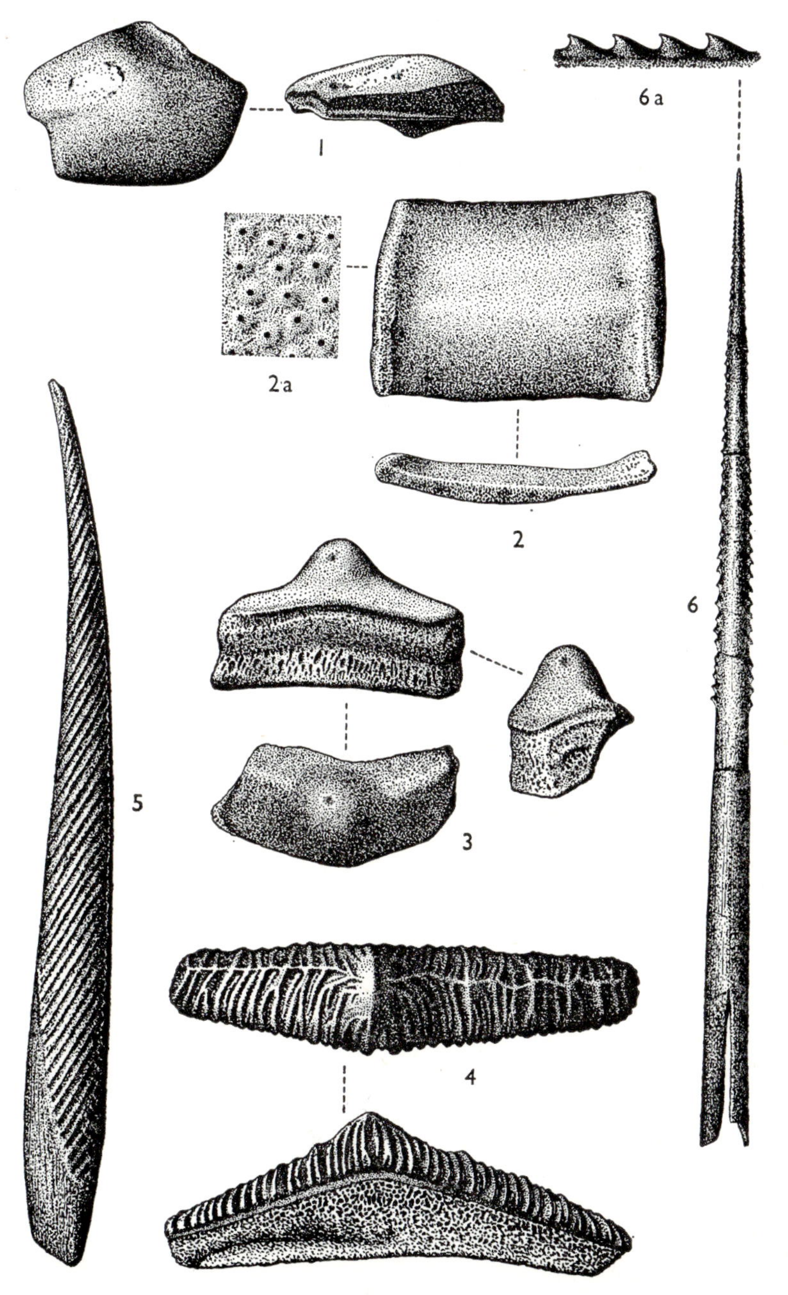

N

Plate 65

Carboniferous Fishes

1, 2. **Megalichthys hibberti** Agassiz. 1, skull (×¼); dorsal view above, ventral view below. 2, restoration of skull in lateral view (×⅙), (after Moy-Thomas). Upper Carboniferous; near Wakefield, West Yorkshire. RANGE: Genus, Carboniferous; Species, Upper Carboniferous. [Syn., *Megalichthys clackmannensis* (Fleming)].

3. **Rhabdoderma tingleyense** (Davis). Gular (throat) plate (×½.) Upper Carboniferous; Tingley, near Wakefield, West Yorkshire. RANGE: Genus, Carboniferous; Species, Upper Carboniferous.

4.* **Rhizodus hibberti** (Agassiz). Tooth (×½.) *a*, section. Lower Carboniferous; Lochgelly, Fife. RANGE: Genus, Carboniferous; Species, Lower Carboniferous.

Plate 65

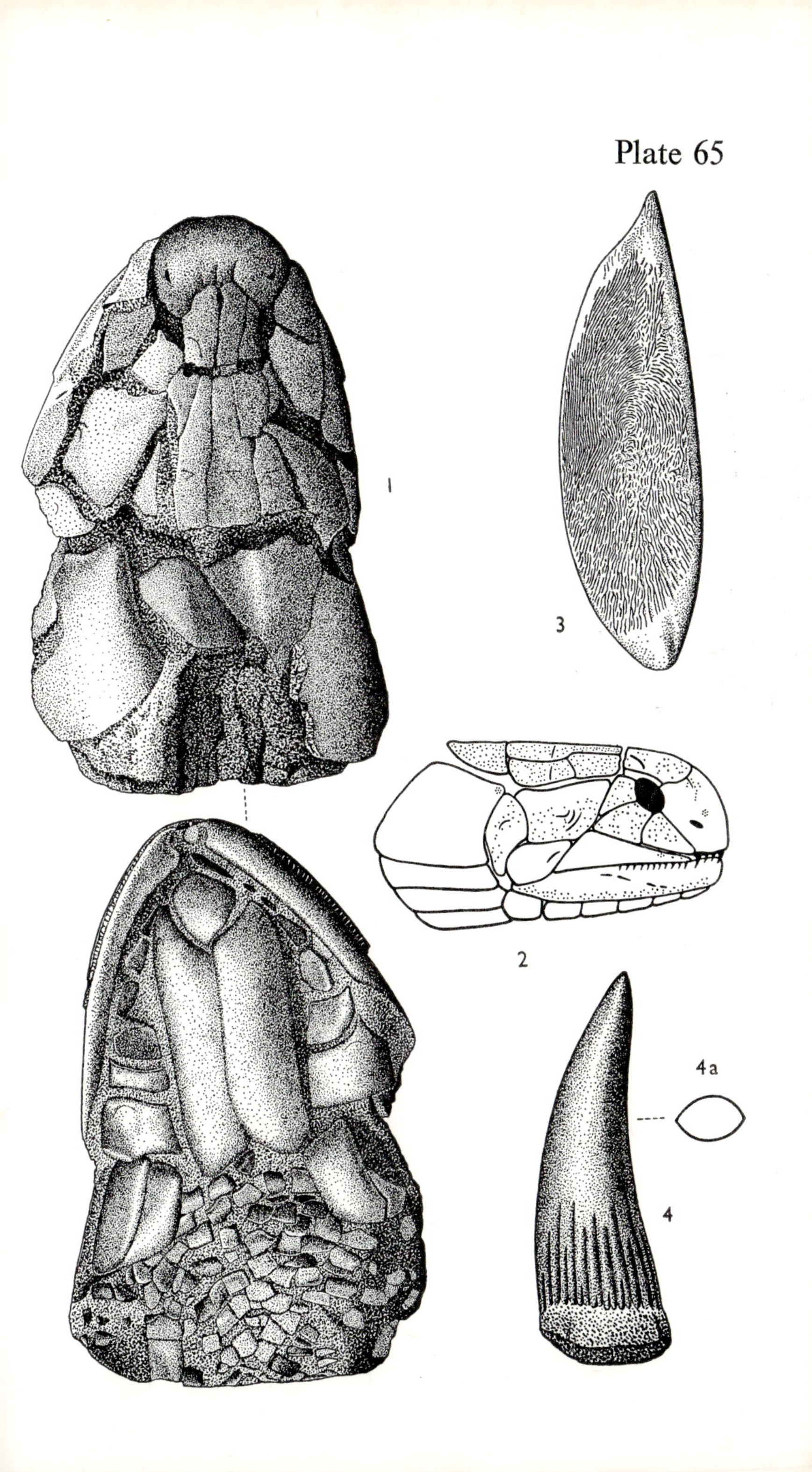

Plate 66

Carboniferous Fish (Fig. 1) and Amphibians (Figs. 2–4)

1. **Sagenodus inaequalis** Owen. Palato-pterygoid with dental plate ($\times\frac{3}{4}$.) Upper Carboniferous, Coal Measures; Newsham, Newcastle. RANGE: Genus, Carboniferous; Species, Upper Carboniferous.

2.* Anthracosaurian vertebra ($\times\frac{1}{2}$.) Upper Carboniferous, Coal Measures; Lowmoor, Bradford, West Yorkshire. RANGE: Upper Carboniferous.

3. **Keraterpeton galvani** Huxley. ($\times\frac{1}{2}$.) Upper Carboniferous, Middle Coal Measures; Jarrow Colliery, near Kilkenny, Ireland. RANGE: Middle Coal Measures.

4. **Megalocephalus** cf. **macromma** Barkas. Skull ($\times\frac{1}{4}$.) Upper Carboniferous, Coal Measures, *modiolaris* Zone; Dawley, Salop. RANGE: Genus, Coal Measures; Species, Ammanian–Morganian Stages. [Syn., *Loxomma allmani* Huxley.]

Plate 66

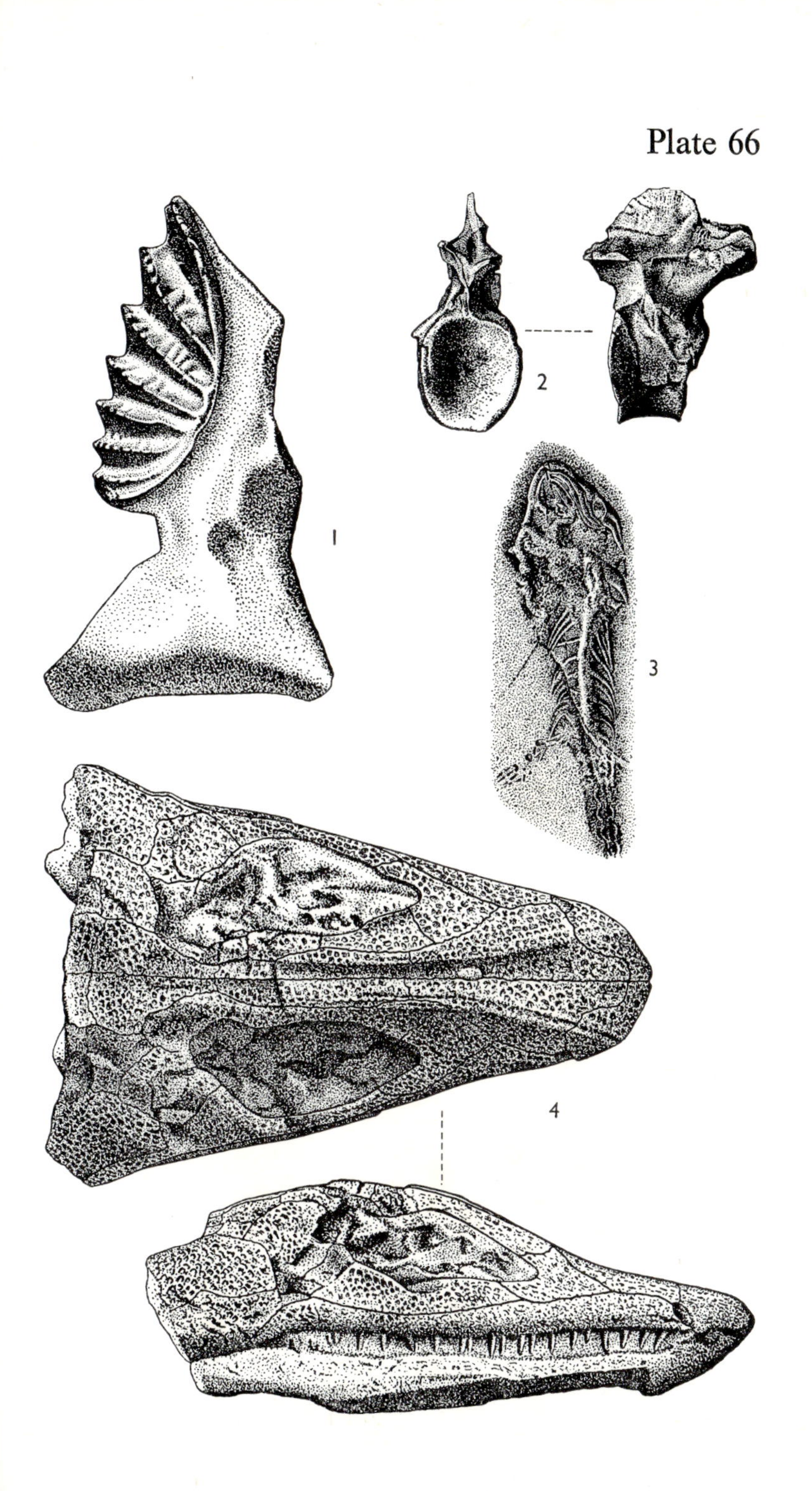

Plate 67

Permian Brachiopods (Figs. 1–4), Foraminifer (Figs. 5, 6) and Bryozoan (Figs. 7, 8)

1, 2.* **Orthothrix excavata** (Geinitz). 1 (×1); *a*, spine bases (×5.) 2, valve with spines (×2.) Magnesian Limestone; Humbleton Hill, near Sunderland. RANGE: Upper Permian. [Syn., *Orthis excavata*, *Strophalosia excavata*.]

3, 4.* **Stenoscisma humbletonensis** (Howse). Magnesian Limestone. 3 (×1.) Humbleton Hill, near Sunderland. 4 (×$\frac{3}{4}$.) Blackhall Colliery Sinking, near Hartlepool, Cleveland. RANGE: Genus, Lower Carboniferous–Permian; Species, Upper Permian. [Syn., *Camarophoria humbletonensis*, *C. multiplicata* King.]

5, 6. **Nodosinella digitata** Brady. 5, exterior (×30.) 6, section (×15.) Magnesian Limestone; Tunstall Hill, near Sunderland. RANGE: Upper Permian.

7, 8.* **Fenestella retiformis** (Schlotheim). Magnesian Limestone. 7 (×1.) Humbleton Hill, near Sunderland. 8 (×2.) East Thickley, near Bishop Auckland, Co. Durham. RANGE: Genus, Ordovician–Permian; Species, Permian.

Plate 67

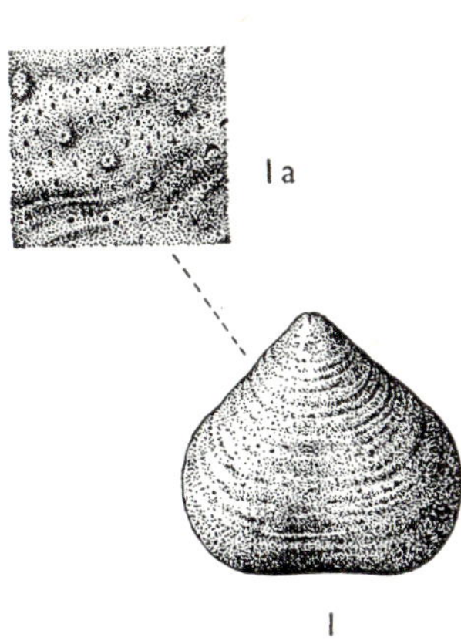

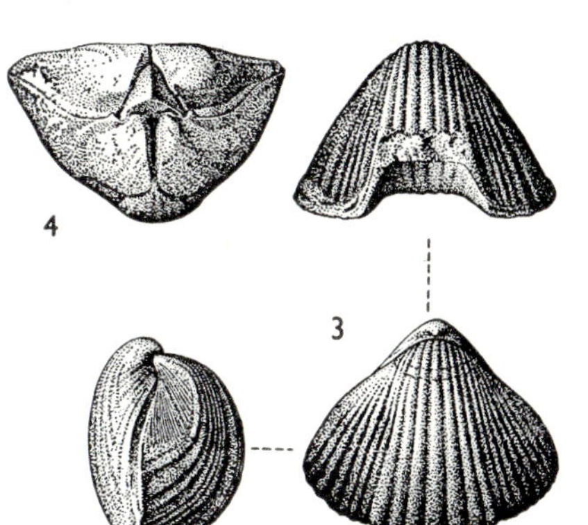

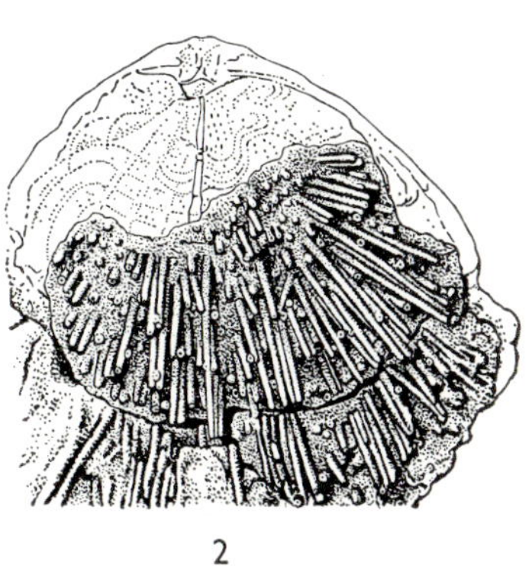

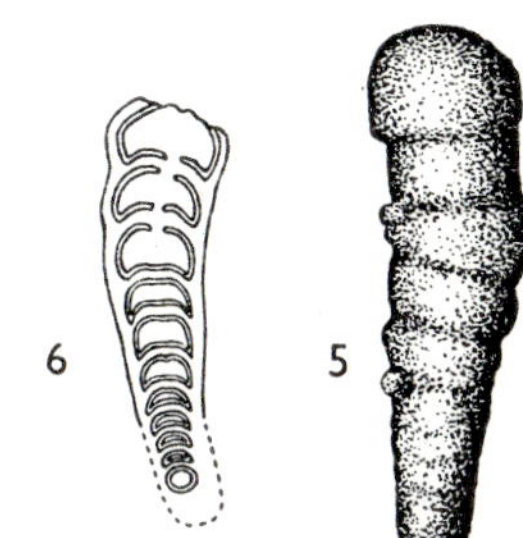

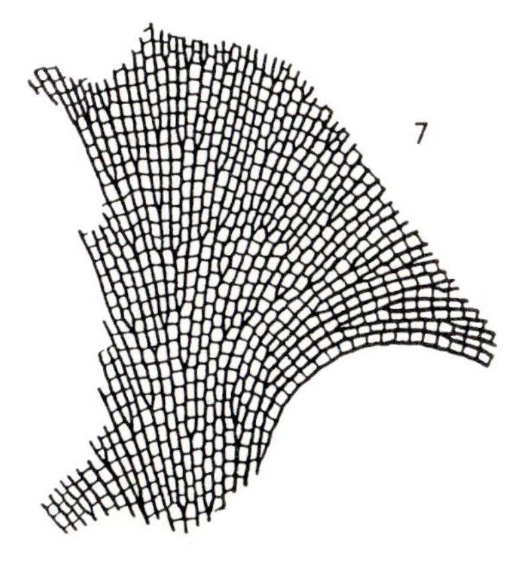

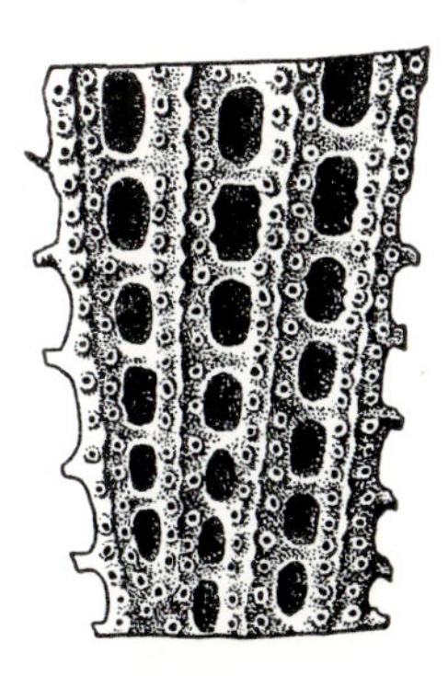

Plate 68

Permian Brachiopods

1.* **Pterospirifer alatus** (Schlotheim). (×1.) Magnesian Limestone; Humbleton Hill, near Sunderland. RANGE: Genus, Devonian–Permian; Species, Upper Permian. [Syn., *Spirifer alatus.*]

2.* **Dielasma elongatum** (Schlotheim). (×2.) Magnesian Limestone; Sunderland. RANGE: Genus, Carboniferous, Viséan Stage–Permian; Species, Upper Permian. [Syn., *Terebratula elongata.*]

3. **Spiriferellina cristata** (Schlotheim). (×3.) Magnesian Limestone; Humbleton Hill, near Sunderland. RANGE: Genus, Devonian–Permian; Species, Upper Permian. [Syn., *Spiriferina cristata.*]

4, 5.* **Horridonia horrida** (J. Sowerby). (×1.) Magnesian Limestone. 4, ventral valve, near Bishop Auckland, Co. Durham. 5, mould of interior of ventral valve, Humbleton Hill, near Sunderland. RANGE: Genus, Permian; Species, Upper Permian. [Syn. *Productus horridus.*]

Plate 68

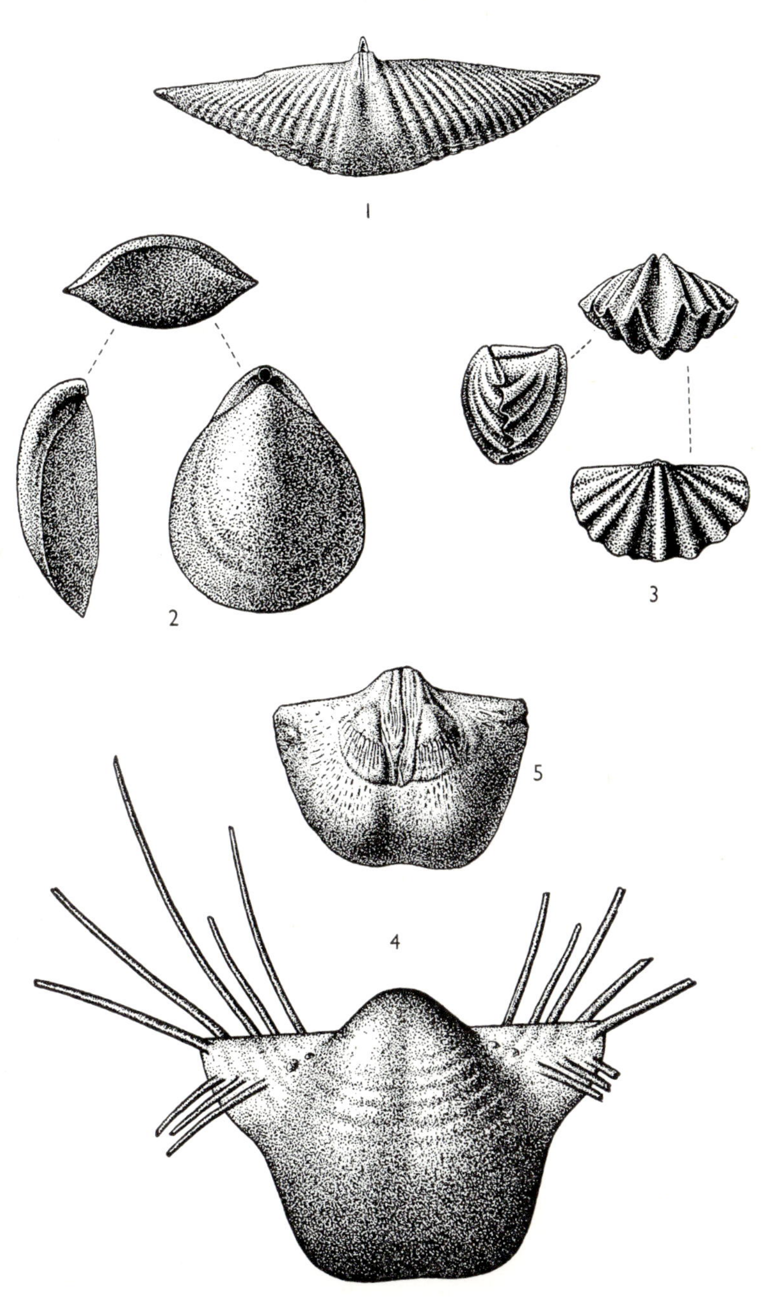

Plate 69

Permian Bivalves (Figs. 1–6) and Fish (Fig. 7)

1. **Parallelodon striatus** (Schlotheim). (×1½). Magnesian Limestone; near Hartlepool, Cleveland. RANGE: Genus, Devonian–Jurassic; Species, Permian. [Syn., *Byssoarca striata*.]

2.* **Permophorus costatus** (Brown). (×1½.) Magnesian Limestone; Ford Lime Quarry, south of Claxheugh Rock, Sunderland. RANGE: Permian. [Syn., *Pleurophorus costatus*.]

3, 4.* **Bakevillia binneyi** (Brown). (×2.) 4, internal mould. Magnesian Limestone; Co. Durham. RANGE: Genus, Permian–Cretaceous; Species, Permian.

5.* **Schizodus obscurus** (J. Sowerby). (×¾.) Magnesian Limestone; Garforth Cliff, near Leeds. RANGE: Genus, Carboniferous–Permian; Species, Permian.

6.* **Pseudomonotis speluncularia** (Schlotheim). (×2.) Magnesian Limestone; Sunderland. RANGE: Permian.

7.* **Palaeoniscus freiselebenensis** Blainville, Tail (×1.) Marl Slate; Ferry Hill, south of Durham City. RANGE: Genus and Species, Upper Permian.

Plate 69

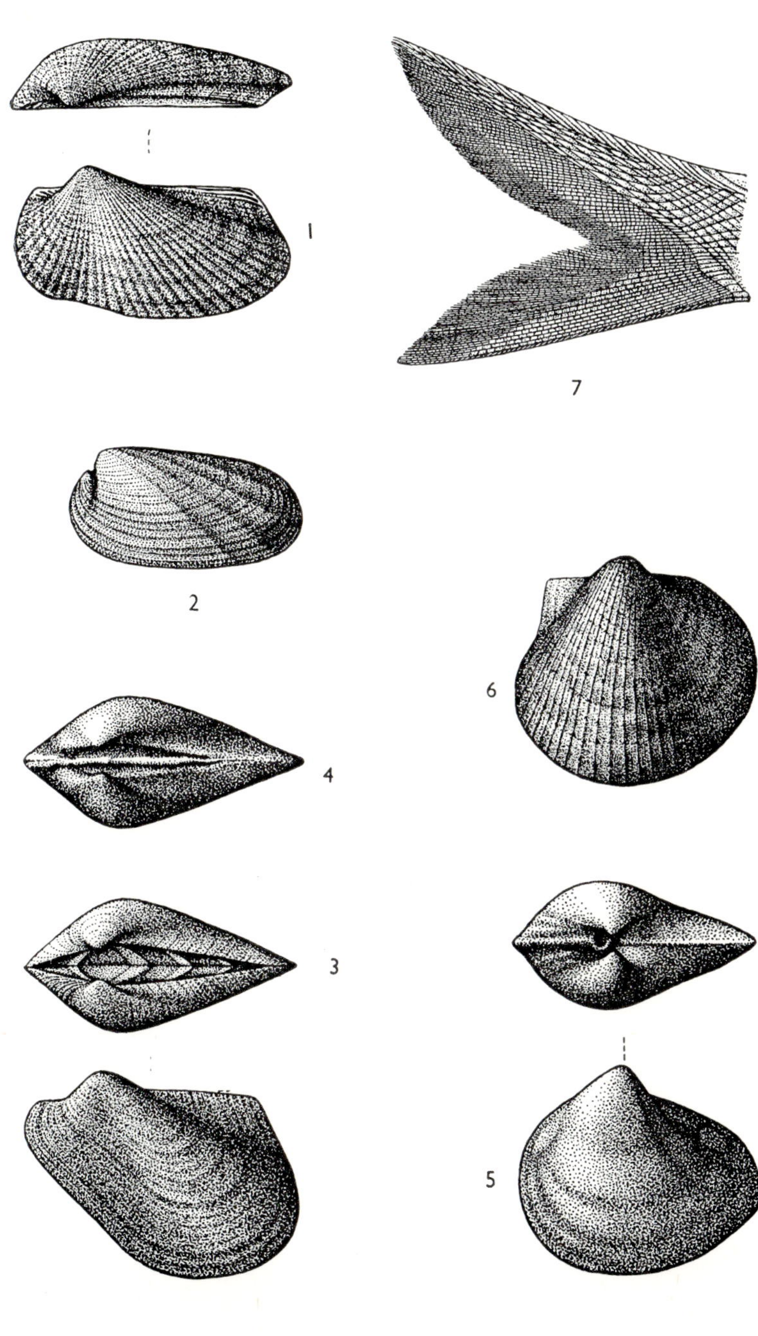

Bibliography

This bibliography is in four parts. The first part lists the publications of the British Museum (Natural History) dealing with Palaeozoic fossils from Britain. They are chiefly in the Bulletin Series (Geology). The second part lists monographs on Palaeozoic fossils published by the Palaeontographical Society. The third part lists the parts of the Treatise on Invertebrate Palaeontology, which aims to give at least one picture of each fossil genus. The fourth part is a small list of works selected from the many hundreds available which describe or revise selected British Palaeozoic fossils. All will be found in most geological libraries, and can be obtained through the inter-library loan system of any public library.

BRITISH MUSEUM (NATURAL HISTORY) PUBLICATIONS

ALDRIDGE, R. J. 1972. Llandovery Conodonts from the Welsh Borderland. *Bull.* **22,** 2.

BALL, H. W., DINELEY, D. L. & WHITE, E. J. 1961. The Old Red Sandstone of Brown Clee Hill and the adjacent area *Bull.* **5,** 7.

BATES, D. E. B. 1968. The Lower Palaeozoic Brachiopod and Trilobite faunas of Anglesey. *Bull.* **16,** 4.

BATES, D. E. B. 1969. Some Early Arenig Brachiopods and Trilobites from Wales. *Bull.* **18,** 1.

BATHER, F. A. 1899. The Genera and Species of Blastoidea, with a list of the specimens in the British Museum (Natural History).

BRUNTON, C. H. C. 1966. Silicified Productoids from the Visean of County Fermanagh. *Bull.* **12,** 5.

BRUNTON, C. H. C. 1968. Silicified Brachiopods from the Visean of County Fermanagh II. *Bull.* **16,** 1.

BRUNTON, C. H. C. 1972. The Shell Structure of Chonetacean Brachiopods and their ancestors. *Bull.* **21,** 1.

COCKS, L. R. M. 1968. Some Strophomenacean Brachiopods from the British Lower Silurian. *Bull.* **15,** 6.

COCKS, L. R. M. 1970. Silurian Brachiopods of the superfamily Plectambonitacea. *Bull.* **19,** 4.

DEAN, W. T. 1959–63. The Ordovician Trilobite faunas of south Shropshire. *Bulls.* **4,** 4; **5,** 8; **7,** 8 and **9,** 2.

DEAN, W. T. 1962. The Trilobites of the Caradoc Series in the Cross Fell Inlier of Northern England. *Bull.* **7,** 3.

ELLIOTT, G. F. 1972. Lower Palaeozoic Green Algae from Southern Scotland and their evolutionary significance. *Bull.* **22,** 4.

ETHERIDGE, R. & CARPENTER, P. H. 1886. Catalogue of the Blastoidea in the British Museum (Natural History).

GARDINER, B. G. 1967. Further notes on Palaeoniscid Fishes with a classification of the Chondrostei. *Bull.* **14,** 5.

HUGHES, C. P. 1969–71. The Ordovician Trilobite faunas of the Builth–Llandrindod Inlier, Central Wales. *Bulls.* **18,** 3; **20,** 4.

JEFFERIES, R. P. S. 1968. The subphyllum Calcichordata (Jefferies, 1967). Primitive fossil Chordates with Echinoderm affinities. *Bull.* **16,** 6.

KALJO, D & KLAAMAN, E. 1965. The fauna of the Portrane Limestone. The Corals. *Bull.* **10, 11.**

MACKINNON, D. I. 1974. The shell structure of Spiriferide Brachiopoda. *Bull.* **25,** 3.

MILES, R. S. 1973. Articulated Acanthodian fishes from the Old Red Sandstone of England, with a review of the structure and evolution of the Acanthodian shoulder-girdle. *Bull.* **24,** 2.

MUIR-WOOD, H. M. 1955. A history of the classification of the Phylum Brachiopoda.

MUIR-WOOD, H. M. 1962. On the morphology and classification of the brachiopod Suborder Chonetoidea.

OAKLEY, K. P. 1966. Some pearl-bearing Ceramoporidae (Polyzoa). *Bull.* **14,** 1.

OWEN, D. E. 1965. Silurian Polyzoa from Benthall Edge, Shropshire. *Bull.* **10,** 4.

PAUL, C. R. C. 1967. The British Silurian Cystoids. *Bull.* **13,** 6.

PRENTICE, J. E. 1967. Lower Carboniferous Trilobites of North Devon and related species from Northern England. *Bull.* **14,** 6.

ROSS, J. R. P. 1966. The fauna of the Portrane Limestone IV. Bryozoa *Bull.* **12,** 3.

SCRUTTON, C. T. 1968. Colonial Phillipsastraeidae from the Devonian of South-east Devon, England. *Bull.* **15,** 5.

SELWOOD, E. B. 1965. Dechenellid Trilobites from the British Middle Devonian. *Bull.* **10,** 9.

SELWOOD, E. B. 1966. Thysanopeltidae (Trilobita) from the British Devonian. *Bull.* **13,** 3.

STENSIO, E. A. 1932. The Cephalaspids of Great Britain.

STRACHAN, I. A redescription of W. Carruthers Type Graptolites. *Bull.* **17,** 4.

TAVERNER-SMITH, R. 1973. Fenestrate Bryozoa from the Visean of County Fermanagh, Ireland. *Bull.* **23,** 7.

TEMPLE, J. T. 1969. Lower Llandovery, (Silurian) Trilobites from Keisley, Westmoreland. *Bull.* **18,** 6.

TOGHILL, P. 1970. Highest Ordovician (Hartfell Shales) Graptolite faunas from the Moffat Area, Southern Scotland. *Bull.* **19,** 1.

WHITE, E. I. 1965. The head of *Dipterus valenciennesi* Sedgwick and Murchison. *Bull.* **11,** 1.

WHITTARD, W. F. 1955. Cyclopygid Trilobites from Girvan and a note on *Bohemilla. Bull.* **1,** 10.

WHITTINGTON, H. B. 1966. Trilobites of the Henllan Ash, Arenig Series, Merioneth. *Bull.* **11,** 10.

WILLIAMS, A. 1963. The Caradocian brachiopod faunas of the Bala District, Merionethshire. *Bull.* **8,** 7.

WILLIAMS, A. 1974. Ordovician Brachiopoda from the Shelve District, Shropshire. *Suppl.* 11.

WOODWARD, A. S. 1889–1901. Catalogue of the Fossil Fishes in the British Museum (Natural History). 4 Vols.

PALAEONTOGRAPHICAL SOCIETY MONOGRAPHS

BASSETT, M. G. 1970– . *The articulate brachiopods from the Wenlock Series of the Welsh Borderland and South Wales.* (In course of publication).

BATTEN, R. L. 1966. *The Lower Carboniferous gastropod fauna from the Hotwells Limestone of Compton Martin, Somerset.*

BULMAN, O. M. B. 1927–1966. *The British Dendroid graptolites.*

BULMAN, O. M. B. 1944–1947. *A Monograph of the Caradoc (Balclatchie) Graptolites from limestones in Laggan Burn, Ayrshire.*

DAVIDSON, T. 1851–1885. *A Monograph of the British Fossil Brachiopoda.*

DEAN, W. T. 1971– . *The Trilobites of the Chair of Kildare Limestone (Upper Ordovician) of eastern Ireland.* (In course of publication).

EDWARDS, H. M. & HAIME, J. 1850–1855. *A Monograph of the British Fossil Corals.*

ELLES, G. L. & WOOD, E. M. R. 1901–1918. *A Monograph of British Graptolites.*

FOORDE, A. H. 1897–1903. *Monograph of the Carboniferous Cephalopoda of Ireland.*

HILL, D. 1938–1941. *A Monograph of the Carboniferous Rugose Corals of Scotland.*

HIND, W. 1894–1896. *A Monograph on* Carbonicola, Anthracomya *and* Naiadites.

HIND, W. 1896–1905. *A Monograph of the British Carboniferous Lamellibranchiata.*

HINDE, G. J. 1887–1912. *A Monograph of the British Fossil Sponges.*

HUTT, J. E. 1974–1975. *The Llandovery Graptolites of the English Lake District.*

INGHAM, J. K. 1970– . *The Upper Ordovician Trilobites from the Cautley and Dent Districts of Westmoreland and Yorkshire.* (In course of publication).

JONES, T. R. 1863. *A Monograph of the Fossil Estheriae.*

JONES, T. R., KIRKBY, J. W. & BRADY, G. S. 1874–1884. *A Monograph of the British Fossil Bivalved Entomostraca from the Carboniferous Formations.*

JONES, T. R. & WOODWARD, H. 1888–1889. *A Monograph of the British Palaeozoic Phyllopoda (Phyllocarida, Packard).*

KING, W. 1850. *A Monograph of the Permian Fossils of England.*

LAKE, P. 1906–1946. *A Monograph of the Cambrian Trilobites.*

LANE, P. D. 1971. *British Cheiruridae (Trilobita).*

LISTER, T. R. 1970– . *The Acritarchs and Chitinozoa from the Wenlock and Ludlow Series of the Ludlow and Millichope area. Shropshire.* (In course of publication).

MILES, R. S. 1968. *The Old Red Sandstone Antiarchs of Scotland: family Bothriolepididae.*

NICHOLSON, H. A. 1886–1892. *A Monograph of the British Stromatoporoids.*

OWENS, R. M. 1973. *British Ordovicain and Silurian Proetidae (Trilobita).*

PAUL, C. R. C. 1973– . *British Ordovician Cystoids.* (In course of publication).

POCOCK, R. I. 1911. *A Monograph of the Terrestrial Carboniferous Arachnida of Great Britain.*

POWRIE, J., LANKESTER, E. R. & TRAQUAIR, R. H. 1868–1914. *A Monograph of the Fishes of the Old Red Sandstone of Britain.*

REED, F. R. C. 1903–1935. *The Lower Palaeozoic Trilobites of the Girvan District, Ayrshire.*

REED. F. R. C. 1920–1921. *A Monograph of the British Ordovician and Silurian Bellerophontacea.*

RICKARDS, R. B. 1970. *The Llandovery (Silurian) Graptolites of the Howgill Fells, Northern England.*

RUSHTON, A. W. A. 1966. *The Cambrian Trilobites from the Purley Shales of Warwickshire.*
SALTER, J. W. 1864–1883. *A Monograph of the British Trilobites from the Cambrian, Silurian and Devonian Formations.*
SLATER, I. L. 1907. *A Monograph of British Conulariae.*
SPENCER, W. K. 1914–1965. *A Monograph of the British Palaeozoic Asterozoa.*
STRACHAN, I. 1971. *A synoptic supplement to* A Monograph of British Graptolites *by Miss G. L. Elles and Miss E. M. R. Wood.*
TEMPLE, J. T. 1968. *The Lower Llandovery (Silurian) brachiopods from Keisley, Westmoreland.*
TEMPLE, J. T. 1970. *The Lower Llandovery Brachiopods and Trilobites from Ffridd Mathrafal, near Meifod, Montgomeryshire.*
TRAQUAIR, R. H. 1877–1914. *The Ganoid Fishes of the British Carboniferous Formations.*
TRUEMAN, A. E. & WEIR, J. 1946–1968. *A Monograph of British Carboniferous Non-marine Lamellibranchia.*
WHIDBORNE, G. F. 1889–1907. *A Monograph of the Devonian fauna of the south of England.*
WHITTARD, W. F. 1955–1967. *The Ordovician Trilobites of the Shelve Inlier, West Shropshire.*
WHITTINGTON, H. B. 1950. *A Monograph of the British Trilobites of the Family Harpidae.*
WHITTINGTON, H. B. 1962–1968. *A Monograph of the Ordovician Trilobites of the Bala Area, Merioneth.*
WOODWARD, H. 1866–1878. *A Monograph of the British Fossil Crustacea belonging to the Order Merostomata.*
WOODWARD, H. 1883–1884. *A Monograph of the British Carboniferous Trilobites.*
WRIGHT, J. 1950–1960. *A Monograph on the British Carboniferous Crinoidea.*

TREATISE ON INVERTEBRATE PALAEONTOLOGY

This series provides pictures of all invertebrate fossil genera and is widely available in geological reference libraries in Great Britain and elsewhere. The series is published jointly by the University of Kansas Press and the Geological Society of America and can be bought from the latter.

Part A. Introduction (not yet published.)
Part B. Protista 1 (Chrysomonadida, Coccolithophorida, Charophyta, Diatomacea, etc.), (not yet published).
Part C. Protista 2 (Sarcodina, chiefly 'Thecamoebians' and Foraminiferida), 1964, xxxi + 900 pp.
Part D. Protista 3 (chiefly Radiolaria, Tintinnina), 1954, xii + 195 pp.
Part E. Second edition. Archaeocyatha, Porifera, 1972 xxx + 158 pp.
Part F. Coelenterata 1956 xvii + 498 pp.
Part G. Bryozoa, 1965 xii + 253 pp.
Part H. Brachiopoda, 1965 xxxii + 927 pp.
Part I. Mollusca (General features, Minor groups, Gastropoda-part), 1960 xxiii + 351 pp.

Part J. Mollusca 2 (Gastropoda), (not yet published).
Part K. Mollusca 3 (Cephalopoda, less Ammonoidea and Coleoidea), 1964 xxviii + 519 pp.
Part L. Mollusca 4 (Ammonoidea), 1957 xxii + 490 pp.
Part M. Mollusca 5 (Coleoidea), (not yet published).
Part N. Mollusca 6 (Bivalvia) 3 volumes, 1969, 1971 xxxviii + 952 pp.
Part O. Arthropoda 1 (Arthropoda general features, Protarthropoda, Euarthropoda General Features, Trilobitomorpha), 1959 xix + 560 pp.
Part P. Arthropoda 2 (Chelicerata, Pycnogonida, Palaeosopus), 1955 xvii + 181 pp.
Part Q. Arthropoda 3 (Crustacea, Ostracoda), 1961 xxiii + 442 pp.
Part R. Arthropoda 4 (Crustacea exclusive of Ostracoda, Myriapoda, Hexapoda) 2 volumes, 1969 xxxvi + 651 pp.
Part S. Echinodermata 1 (Echinodermata general features, Homalozoa, Crinozoa, exclusive of Crinoidea), 1968, xxx + 650 pp.
Part T. Echinodermata 2 (Crinoidea), (not yet published).
Part U. Echinodermata 3 (Asterozoans, Echinozoans), 1966 xxx + 695 pp.
Part V. Second edition. Graptolithina, 1970 xxxii + 163 pp.
Part W. Miscellanea (Conodonts, Conoidal Shells of uncertain affinities, Worms, Trace Fossils, Problimatica), 1962 xxv + 259 pp.
Part X. Addenda, Index. (not yet published).

OTHER SELECTED PUBLICATIONS

ANDREWS, H. N. 1961. *Studies in Paleobotany*. New York and London.
BAKER, E. W. & WHARTON, G. W. 1952. *An Introduction to Acarology*. New York.
BATHER, F. A. 1890–1892. British Fossil Crinoids, Pts I–VIII (scattered through) *Ann. Mag. Nat. Hist.*, London (6) **5–7, 9.** Chiefly Silurian.
BISAT, W. S. 1924. The Carboniferous goniates of the north of England and their zones. *Proc. Yorks. Geol.* Soc., **20** : 40–124.
CROOKALL, R. M. 1929. *Coal Measure Plants*. London.
HALLAM, A. 1973. Distributional Patterns in Contemporary Terrestrial and Marine Animals. *Special Papers in Palaeontology*. No. 12, 93–105.
HUGHES, N. F. 1973. Organisms and Continents through Time. *Special Papers in Palaeontology*, No. 12.
KIDSTON, R. 1923–1925. Fossil Plants of the Carboniferous rocks of Great Britain. *Mem. Geol. Surv. Gt. Brit. Palaeontology*, **2,** 1–6.
LANG, W. D. & SMITH, S. 1927. A Critical revision of the Rugose Corals described by W. Lonsdale in Murchison's *Silurian System. Quart. J. Geol. Soc. Lond.*, **83,** 448–491.
MOY-THOMAS, J. A. & MILES, R. S. 1971. *Palaeozoic Fishes* 2nd. edition. London.
MUIR-WOOD, H. M. & COOPER, G. A. 1960. Morphology, classification and life habits of the Productiodea (Brachiopoda). *Mem. Geol. Soc. Amer.*, **81**.
PETRUNKEVITCH, A. 1953. Palaeozoic and Mesozoic Arachnida of Europe. *Mem. Geol. Soc. Amer.*, **53**.
SMITH, A. H. V. & BUTTERWORTH, M. A. 1967. Miospores in the Coal Seams of the Carboniferous of Great Britain. *Special Papers in Palaeontology*, No. 1.

WALTON, J. 1933. *An Introduction to the study of the fossil plants.* London.
WILLIAMS, A. 1962. The Barr and Lower Ardmillan Series (Caradoc) of the Girvan District, south-west Ayrshire, with descriptions of the Brachiopoda. *Mem. Geol. Soc. Lond.*, **3.**
WILLS, L. J. 1959–1960. The external anatomy of some Carboniferous 'Scorpions'. *Palaeontology London*, **1** : 261–283; **3** : 267–332.

Index

Names of species in current use are printed in **heavy** type. Synonyms, or discarded names, are in *italics*. In the plate references, the figure in **heavy** type **(1)** refers to the plate; that in ordinary type (1) to the figure.